MARKETING FASHION
FOOTWEAR

Bloomsbury Visual Arts

An imprint of Bloomsbury Publishing Plc

Imprint previously known as AVA Publishing

50 Bedford Square   1385 Broadway
London   New York
WC1B 3DP   NY 10018
UK   USA

www.bloomsbury.com

**Bloomsbury Visual Arts, BLOOMSBURY and the Diana logo are
trademarks of Bloomsbury Publishing Plc**

British Library Cataloguing-in-Publication Data
A catalogue record for this book is available from the British Library.

ISBN: PB: 978-1-4725-7931-7
   ePDF: 978-1-4725-7932-4

Library of Congress Cataloging-in-Publication Data
Armstrong-Gibbs, Fiona.
Marketing fashion footwear: the business of shoes / Fiona Armstrong-Gibbs
and Tamsin McLaren.
pages cm
Includes bibliographical references and index.
ISBN 978-1-4725-7931-7 (pbk.) -- ISBN 978-1-4725-7932-4 (epdf) --
1. Footwear industry. 2. Fashion merchandising. 3. Shoes--Marketing.
I. McLaren, Tamsin. II. Title.
HD9787.A2A76 2016
685'.3100688--dc23
2015016841 LC record available at http://lccn.loc.gov/2015040645

Series: Required Reading Range

Cover design: Louise Dugdale
Cover image: courtesy Joanne Stoker

Typeset by Roger Fawcett-Tang
Printed and bound in China

MARKETING FASHION
FOOTWEAR

THE BUSINESS OF SHOES

Fiona Armstrong-Gibbs and Tamsin McLaren

Bloomsbury Visual Arts
An imprint of Bloomsbury Publishing PLC

B L O O M S B U R Y
LONDON · OXFORD · NEW YORK · NEW DELHI · SYDNEY

CONTENTS

01

THE FASHION FOOTWEAR CONSUMER

02

FOOTWEAR DESIGN, CONSTRUCTION, AND PRODUCTION

06

VISUAL MERCHANDISING AND DESIGN CONCEPTS FOR RETAIL, E-TAIL & WHOLESALE

07

BRAND IDENTITY AND PROTECTION

THE GLOBAL FOOTWEAR TRADE

THE RETAIL AND E-TAIL LANDSCAPE

MANAGEMENT STRATEGIES FOR RETAIL GROWTH

BRAND MANAGEMENT

MARKETING COMMUNICATIONS

FOREWORD

The footwear industry has seen a number of quantum shifts in the last decade—not just from the perspective of continued growth, but also in how customers consume media and engage with brands. This has been stimulated in part by the evolution of technology that has democratized the media landscape. Brands can now reach their customers by creating a dialogue across multiple platforms.

There is also the critical challenge of understanding channels of distribution that have exploded in recent years through the onset of omni-channel retailing. Both retailers and brands alike have been quick to realize this market growth and potential, by opening up increasingly sophisticated offers to the consumer across digital and brick-and-mortar channels.

Understanding this marketplace in order to drive, reach, and stimulate awareness and demand to resonate with informed customers has become increasingly complex. Historically there has not been a single book that addresses the critical strategic issues that face any individual or organization on how to market and distribute its products.

This book provides informative and practical insight into how the whole journey from concept to customer is becoming more tailored toward increasingly savvy and demanding consumers within this highly competitive environment.

I would have loved to have read something like this when I first set out in the industry twenty years ago, as it provides a clear and no-nonsense approach on how to understand the business—from production through to season launch and retail, and from a marketing perspective, how to engage with consumers by setting out and reinforcing a clear positioning and reaching them in an authentic and credible way.

Simon Jobson, Global Marketing Director at Airwair International Ltd.—Dr. Martens

PREFACE

We established early on in our teaching careers that there was a gap in academic literature and that no single text specifically addressed the fashion footwear business. Fashion marketing and management books focus predominantly on apparel, and footwear-specific publications address design, production, and historical factors rather than examining the business from a management and marketing perspective.

In many conversations with industry and academic colleagues alike, the feeling was that students were not always aware of opportunities within the sector, and that there was no single "go-to" book for industry professionals either; *Marketing Fashion Footwear* will be useful for students studying fashion marketing, management, buying, merchandising, and design as the aim is to provide a much needed balance and broader perspective on a fast-growing sector. It will also provide industry professionals with a view on how footwear differs from other product areas, and so it can be used to support individual training and development needs.

Marketing Fashion Footwear therefore does not focus on performance footwear, designed for athletic and health and safety end use. The words footwear and shoes may be used interchangeably, but generally industry professionals use the term "footwear."

The chapters have been designed to explore the journey from customer to product and marketing concept throughout its different stages: consumer behavior, research and development, trend forecasting, product creation, the supply chain, retail and wholesale distribution, brand management, and marketing and public relations. Each chapter contains key learning objectives, an introduction to the topic, an Ethics in Action section, case studies, and an interview with a key industry professional.

FASHION FOOTWEAR IS FOOTWEAR THAT HAS BEEN DESIGNED, MANUFACTURED, AND MARKETED WITH A SPECIFIC FOCUS ON FASHION STYLING AND TRENDS. FASHION FOOTWEAR IS BOUGHT BY A CONSUMER WHO CONSIDERS VISUAL APPEAL AND STYLE AS WELL AS SEASONALITY AND FUNCTIONALITY. BRAND AND AESTHETICS CAN TAKE PRIORITY OVER PRACTICALITY, DURABILITY, AND COMFORT.

Key marketing and brand management theories are applied from leading international and upcoming footwear and fashion brands, and within the Ethics in Action sections moral or socially pertinent issues that are facing the industry are highlighted.

Footwear is a vast and dynamic industry, with many interesting and varied job roles requiring a myriad of both transferable and specific skills. Key job roles are discussed within the chapters via industry insights from designers, brand owners, retail buyers and store owners, brand managers, and digital specialists. Detailed figures, diagrams, and images support the text and contextualize theory, and each chapter is summarized and a set of discussion questions and exercises are provided. These can be used as a basis for in-class discussion, and they form the basis for further study and wider research. Finally, the key terms are listed to clarify and support the learning objectives.

We begin by focusing on the consumer (in Chapter 1) and exploring how the fashion footwear consumer displays a variety of behaviors that differ from that of apparel, embedding much more emotion into choosing, wearing, and disposing of shoes. Chapter 2 explores how manufacturing footwear differs from clothing and follows the evolution from traditional to modern and future techniques in design, construction, and production. There is a focus on global hubs of production and regions that specialize in specific production techniques, their development, and place in the manufacturing landscape of the twenty-first century.

We then look at the international nature of the industry, as Chapter 3 assesses the impact of globalization and regional and country trends are discussed to provide an understanding of the history, current challenges, and future directions for the sector. In addition, this chapter looks at the relevance of changing legislation around quotas and tariffs, export and import legislation, and the key role that footwear plays as part of different countries' overall economic stability and growth.

Chapters 4, 5, and 6 outline the evolving retail and e-tail landscape and how this has altered dramatically in the last twenty years. Retailers and brands of all sizes are required to make more innovative products and have better quality production, an improved and more seamless customer experience, as well as more exciting and intriguing marketing and communications —gone are the days when "product is king" alone. Different countries have varying retail distribution models, but the uniting factor in the digital age is the power of omni-channel distribution and marketing.

Chapter 7 identifies issues faced by designers, brands, and retailers around brand identity and protection in today's counterfeit culture. Protecting intellectual property and some of the wider-reaching legal implications of recent cases and decisions, as well as the effect fakes have on the industry as a whole, are discussed. Brand management is explored in depth in Chapter 8 via the discussion of critical success factors for longevity of different types of brands through brand collaborations and an understanding of brand equity and value, as well as the iconic nature of brands and footwear.

The final chapter covers the scope of marketing communications and the channels used to reach key opinion leaders and consumers, both in traditional and new media, and how this affects the brand's message. It assesses the various promotional tools used for successful campaigns in the twenty-first century and considers footwear as a product that has been elevated through the use of product placement and celebrity endorsement.

Extensive primary research was undertaken in the United States, Canada, and Europe to support secondary research, in order to provide a snapshot view of how the international fashion footwear industry works. Due to the length of the research, writing, and publication process there is always scope to update material, and we hope that students and industry professionals will be inspired by the sources made available to them in this book.

Shoes are the subject of fairy tales, films, art, and exhibitions; they hold powerful cultural connotations, and yet the business as a whole and how it operates has not yet been captured in a single publication—until now. For those studying or working with footwear and fashion, understanding the business of shoes is essential. The last twenty years have seen the product category grow exponentially as a key fashion item, outperforming other sectors in the industry. And increasingly the fashion industry is recognizing opportunities within this product category, and so the focus, investment, and employment opportunities are growing.

We hope you find this book to be useful and informative!

Fiona Armstrong-Gibbs and Tamsin McLaren

01

THE FASHION FOOTWEAR CONSUMER

Learning Objectives

- Outline global footwear consumption patterns and define consumer behavior with reference to demographics, psychographics, and social factors.
- Examine customer motivations and issues of self-concept and how this impacts on the decision-making process and disposal of shoes.
- Identify the key research tools to develop a specific understanding of the fashion footwear consumer.
- Outline the fashion cycle and the key aspects of trend forecasting and analysis for footwear.

1.1 Fashion style tribe at London fashion week
Connecting and understanding key customer groups is essential when developing a successful marketing strategy.

INTRODUCTION

Consumer behavior is the study of how individuals make decisions about how to spend their available resources, time, money, and effort when purchasing products. This chapter will cover the key concepts underpinning the behavior consumers display when searching for, purchasing, wearing, evaluating, and disposing of footwear. It is important to make the distinction between the customer who is the purchaser of shoes—either an individual or a retailer buyer—and the consumer who is the end user or wearer; however these may be used interchangeably. Embedded in behavior is the evolution of trends and process that consumers adopt and adapt to fashion change. This chapter will address the behavior of today's footwear shopper with reference to trends, the macro environment, and influencers.

GLOBAL FOOTWEAR CONSUMPTION

Historically, footwear was a functional purchase; most customers bought a black or brown pair of shoes, sandals for summer, boots for winter, and possibly a pair of "Sunday best" or evening shoes for special occasions. Footwear was an investment, expected to be durable and repairable. The 1980s and '90s saw the increasing casualization of clothing, a blending of acceptable styles for work and personal life. People wore comfortable sportswear as casual and leisure wear, adopting the same principle in their footwear and allowing for sneakers and trainers to enter the fashion market. The widening of choice and options in footwear styles and colors gained momentum through the 1990s and firmly established footwear as a commodity item. The "fast fashion" clothing model in the new millennium also contributed to the increase in footwear sales, satisfying consumer demand for affordable and stylish fashion footwear.

The global footwear market grew by 3.9 percent in 2014 to reach a value of $289.7 billion and between 2010 and 2014 saw an annual growth rate of 4.8 percent. Across the world the footwear sector has seen a higher percentage growth than apparel. Athletic and casual footwear has driven this growth alongside increased consumer loyalty to brands, with consumers willing to replace footwear more often than before.

Which Countries are the Largest Consumers of Fashion Footwear?

With the largest economy in the world and a population of over 1.3 billion, China is the world leader in manufacturing, exporting, and consuming footwear. Both India and Brazil have high consumption percentages, however this is reflective of a large population, and much of what is produced in these countries stays in the home market. As these economies emerge and offer their population higher standards of affluence and living, it is predicted that their purchases per capita will increase. The United States, Germany, France, United Kingdom, and Japan are the top five leading markets by volume and by pairs per capita. None of these countries are leading manufacturers;

Country	Pairs of shoes in millions	% of world share	Population in millions	GDP per capita $	Pairs of shoes purchased per head
1. China	3,646	18.80%	1,368	$7,589	3
2. United States	2,295	11.80%	319	$54,597	7
3. India	2,048	10.50%	1,260	$1,672	2
4. Brazil	807	4.20%	203	$11,604	4
5. Japan	607	3.10%	127	$36,332	5
6. Indonesia	548	2.80%	251	$3,534	2
7. United Kingdom	523	2.70%	65	$45,653	8
8. Germany	435	2.20%	81	$47,590	5
9. France	434	2.20%	64	$44,538	7
10. Russian Federation	411	2.10%	144	$12,926	3

however, they have well-established and -developed retail markets. Coupled with high levels of gross domestic product (GDP), this suggests that consumers have the ability to make optional "fashion" purchases. In general, the wealthier the country, the more shoes people buy.

1.2 The world's largest consumers of footwear in 2014
In 2014 over 24 billion pairs of shoes were produced. Developed, affluent countries such as the United States, Japan, and Western Europe are still keen consumers of footwear but have seen their overall consumption patterns decline slightly since the economic crisis of 2008. The last few years have seen Asian countries such as China and Indian increase their consumption of footwear.
Adapted from the World Footwear Yearbook 2015
(http://www.worldfootwear.com)

WHAT IS CONSUMER BEHAVIOR?

Consumer behavior is the actions that individuals, groups, or professional organizations display when evaluating, selecting, purchasing, using, and disposing of goods and services and how these satisfy their needs and impact wider society.

Fashion footwear for many people is now a discretionary purchase; with increased brand awareness, innovative technology, and huge choice, retailers operate in a highly competitive and sophisticated environment. Their primary goal is to recognize and maintain their customers.

1.3 Customer segmentation
In order to understand the consumer it is necessary to segment, or split, them into groups based on similarities such as age and gender, as this will have an influence on what they buy and how they shop. Many young people are heavily influenced by the brands that their friends and peers are wearing.

Identifying and Segmenting the Consumer

Retailers and brands split, or "segment," mass (homogeneous, undifferentiated) markets into heterogeneous markets that share similar characteristics. These markets—or potential customers—are then defined by demographic, psychographic, and social factors. Demographics are objective and quantifiable data such as population, age, and sex and can explain who buys. They form the first part of a consumer profile. Psychological factors such as personality, self-awareness, attitudes, and opinions plus social factors such as culture, class, reference groups, and family life cycle also need to be included and help to explain why people buy. While the geographic and socio-demographic data are reasonably straightforward to gauge, it is much harder to quantify these other behavioral aspects of the consumer.

1.3

1.4

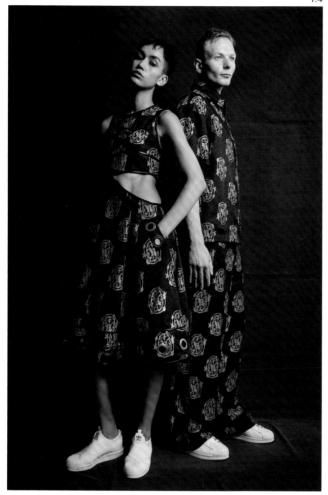

Demographics

Target customers are measured by indicators of gender, age, and ethnicity as well as geographical data such as housing type/location and economic data such as income. These indicators build a pattern over time that may show an aging population, falling birth rates, economic variables, and fluctuations in employment. All of these help to build a picture of who the marketer may wish to target or how it may adapt and develop its products. Online shopping behavior via gender shows that men are more like to shop on intermediary websites such as Amazon for their shoe purchases, whereas women tend to enjoy searching and go to specific retailers' site to browse and purchase.

> **" DEMOGRAPHICS ALLOW US TO DESCRIBE WHO BUYS AND PSYCHOGRAPHICS ALLOWS US TO UNDERSTAND WHY THEY BUY."**
> SOLOMON & RABOLT, 2009

1.4 Unisex styling at Ming Pin Tien
A Chinese designer Ming Pin Tien photo shoot showcases adidas sneakers styling for both men and women as a fashion choice.

1.5 Family life cycle stages
Consumers' priorities often change as their family status evolves; therefore, their footwear preferences and financial spending power may affect their choice of footwear.

1.5

	Household composition or family life stage/cycle*	Footwear choice
1	Bachelor and Peacock (students, low income, working) single	Young, sexy, affordable, and fun
2	Early nesters (working) couples without children	Glamorous, seeking status and variety
3	Full nesters (working) couples with children	Practical, stylish, and comfortable
4	Empty nesters (working or retired) couples without children	Investment, durable, comfortable, and purposeful but stylish
5	Solitary survivors (retired) singles	Comfortable and supportive but presentable

*Also consider single-parent households, parents with grown-up children still living at home, and house shares in large cities, universities, etc.

Psychographics

Consumers are also segmented by psychographic factors, which requires an awareness of a person's values, attitudes, and lifestyle and may explain why people make the choices they do. These factors are critical for developing an accurate picture of the potential and existing customer. Understanding someone's demographic data, such as income or profession, alongside their psychographic profile may help to calculate how much spare time they have and how they use this time. Where they live could also signify how conscious they are of their status or class. Consumer attitudes, however, play a role as well. These enduring evaluations of people, brands, social situations, and issues also may help to explain lifestyle and consumption patterns. Insights into the importance of the purchase and perceived risk an individual may feel when looking for the product are also important. All of these elements of market research can lead marketers to work out a profile of their target customer and gain a better understanding of the attitudes toward the brand or retailer.

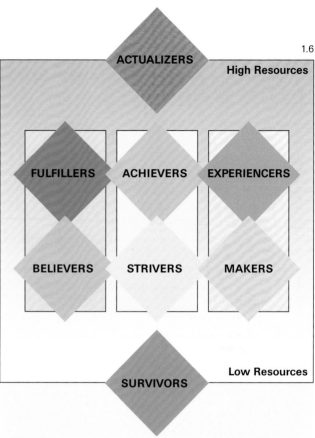

1.6

1.6 VALS 2 Segmentation System
Psychographic research that identifies people's values, attitudes, and lifestyles (VALS) is considered to have surpassed the simplicity of demographic categories, and in the United States many marketers will use the VALS 2 Segmentation System to identify different customer groups.

Actualizers
High income, leaders, self-esteem, image, status, and taste are important

Thinkers
Well educated, mature, responsible, knowledgeable, rational and practical, open minded, and accepting of social change

Achievers
Affluent, motivated by achievement, career oriented, safe, and predictable

Experiencers
Young, affluent, self-expressive, impulsive, and avid spenders

Believers
Modest income, predictable, principled, and dependable with local view and focused on family and community

Strivers
Low/modest income, emulate and aspire, motivated by achievement but limited by resources

Makers
Low/modest income, self-sufficient, practical, and functional

Survivors
Little income, meeting needs, cautious, and loyal

Reference Groups

Our buying decisions are heavily influenced by social reference groups who have a profound effect on how we live our lives and make our purchases. Normative reference groups such as family will influence general or basic codes of behavior, manners, or values. Comparative reference groups, such as peers, classmates, or inaccessible celebrities and opinion leaders, help to form benchmarks and influence how we rate ourselves and what we wear, along with more specific types of attitudes or behavior. An example of this could be how classmates influence a person's choice of branded sneaker. However, in some cultures the role the family plays in an individual's choice may be as important as their comparative reference groups. Bound in this are deep ties to a social class or order they belong to; these may be difficult for an observer to decipher.

Consumers go to different media platforms to gain knowledge, often merging reference groups and commercial advertising and promotional messages through social media such as Facebook, Instagram, and individual fashion blogs. Traditional reference groups are becoming increasingly blurred due to the rise of digital communications and online communities, as well as fashion bloggers who may work with brands but appear to relate to their readers and viewers in a much more sociable and informal way (see Chapter 9 for further discussion).

" PERSONALITY REFERS TO A PERSON'S UNIQUE PSYCHOLOGICAL MAKEUP AND HOW IT (CONSISTENTLY) INFLUENCES HOW A PERSON RESPONDS TO HIS OR HER ENVIRONMENT."
SOLOMON & RABOLT, 2009

" FOR ADOLESCENTS, SHOES ARE A KEY SIGNIFIER OF THEIR IDENTITIES, AND THE SHOES THEY DESIRE OFTEN CONFLICT WHAT THEIR PARENTS REGARD AS APPROPRIATE. SHOES APPEAR AS A KEY VEHICLE THROUGH WHICH ADOLESCENTS AND YOUNG ADULTS WORK OUT ISSUES OF IDENTITY, INDIVIDUALISM, CONFORMITY, LIFESTYLE, GENDER, SEXUALITY, ETHNICITY, AND PERSONALITY."
BELK, 2003

CONSUMER MOTIVATIONS

Motivation is identified as a need that has been sufficiently stimulated to prompt someone to seek satisfaction. When shopping, consumers use the word "need" in different ways. A utilitarian need fits a purpose, such as a pair of boots to keep our feet warm and dry, whereas hedonic needs are subjective and emotional, often relating to self-confidence: we feel the need for a pair of acceptable fashion shoes to help us keep up with the latest trends and find social acceptance.

Needs can be physical or emotional, and both impact when buying a pair of shoes. Psychologist Abraham Maslow (1954) developed a theory to explain the motivational behavior that leads us to consume a variety of products that is still relevant today. Once our basic physiological needs have been met, we move up the hierarchy, satisfying the varying necessities in the process. Social acceptance and self-esteem assurances ultimately lead to self-fulfilment and enriching experiences.

Emotional Purchasing

Emotion is a key driver in discretionary purchases and is related to moods and feelings rather than physical satisfaction. New shoes from an ethical brand such as Toms or Gandys flip-flops may help to repair a bad mood or maintain a good one, making the purchaser feel good about supporting a cause. Emotion is also related to impulse buying. Emotion can impair judgment, relying on cues such as brand labels, low prices, and discounting so the customer requires less information or analysis before making the purchase. When we are emotional it is also easy to avoid negative cues, such as ill-fitting or uncomfortable shoes. Fast fashion footwear is sold in pairs, allowing customers to serve themselves. They may buy quickly without trying on and only when they get home realize they have ill-fitting shoes, two right-footed shoes, or a pair of different sizes.

Self-Concept

Most consumers feel varying degrees of positivity about themselves, known as self-esteem. Fashion and appearance management link with self-esteem; we look after our appearance when we feel good, and we may over-compensate when we feel insecure. We use footwear to transform ourselves, to aspire to an ideal, to make us perform better or escape to a fantasy version. Different shoes help us play out different acts in our life. Converse to relax in, Hunter wellington boots to enjoy a music festival and Christian Louboutin heels for a date night. Legendary 1950s French shoe designer Roger Viver is quoted as saying, "To wear dreams on one's feet is to begin to give a reality to one's dreams, and to nourish the hope that they will bring on other dreams."

The past decade has seen much debate in the media about the ideal body shape, but footwear for many people taps into a different emotion. Footwear is ageless and size-less, and it transcends ethnicity. The right shoes can be empowering, enable us to feel taller, alter our deportment and stride, and add the finishing touches to our personality.

1.7

Self-Actualization
Creativity, Problem
Solving, Authenticity,
Spontaneity

Esteem
Self-Esteem, Confidence, Achievement

Social Needs
Friendship, Family

Safety and Security

Physiological Needs (Survival)
Air, Shelter, Water, Food, Sleep, Sex

" **SELF-CONCEPT REFERS TO THE BELIEFS A PERSON HOLDS ABOUT HIS OR HER ATTRIBUTES AND HOW HE OR SHE EVALUATES THEM.**"
SOLOMON & RABOLT 2009

1.7 Maslow's Hierarchy of Needs
Defining "need" through Maslow's theory allows the marketer to identify at which point the consumer will purchase shoes to fulfill a certain need, for example a pair of rubber wellington boots to keep your feet dry at a muddy music festival.

1.8 Shoe choices reflect elements of our personality
Consumers shop with a view of themselves and buy product that they believe reflects this view both consciously and subconsciously. Fashion footwear choice is about crafting an image of and identity for ourselves.

" WHEN A WOMAN BUYS SHOES, SHE TAKES THEM OUT OF THE BOX AND LOOKS AT HERSELF IN THE MIRROR. BUT SHE ISN'T REALLY LOOKING AT HER SHOES—SHE'S LOOKING AT HERSELF. IF SHE LIKES HERSELF, THEN SHE LIKES THE SHOES."
CHRISTIAN LOUBOUTIN

Conspicuous Consumption—Income versus Image

Sociologist Thorstein Veblen developed the theory of conspicuous consumption in 1899, which suggests that we use certain products such as clothing and accessories to display our financial wealth or value. This is evident when purchasing branded footwear. Logos such as Hunter on boots and Christian Louboutin's red-soled high heels project signs of status in society, wealth, and awareness of fashion trends. We place a large priority on what other people think of our footwear choice, and shoes have the ability to convey status. In the past, high heels particularly represented privilege in part due to their impracticality; women did not walk very far, let alone run or physically exert themselves with manual work. Heels were the domain of the leisure class; today it is the "limo class" of celebrities and socialites. Shoes can symbolize social distinction or acceptance from one's peers, sexual attraction, and a reflection of the current zeitgeist or spirit of the time. Undoubtedly money and material goods enable consumers to develop their sense of identity, feelings, and relationships with others; therefore our choice of footwear may well symbolize how much money we have.

Jasmin Sanya, Footwear Buyer at Harvey Nichols, UK

What do you think motivates people to buy footwear?

Footwear has always been an important piece of any outfit, not least because a pair of shoes is necessary. Many people have a huge affection towards shoes and the effect they can have on one's persona or mood. In our culture, people often judge one another on their choice of shoes, whether it's for a first date or an interview; we can use our choice in footwear to show an extension of our personality and character.

What or who influences the consumer to buy?

In today's digitally driven culture, we are finding a large proportion of our customer base influenced by the images they see on social media; a celebrity following for a brand is prevalent in today's market. The consumer is constantly looking to seek out the next big thing, something fresh and undiscovered, but traditional media channels such as product placement in magazines and advertising campaigns still has its place and impact on the brand's customer.

1.9

> " OURS SOLES ARE THE MIRROR OF OUR SOULS ... THEY SEPARATE US FROM THE DIRT AND SYMBOLIC IMPURITY OF THE GROUND AND ELEVATE OUR STATUS AND GENDERED SENSE OF SELF."
> BELK 2003

1.9 Multiple shoe personalities
The popularity of social media channels such as Instagram allows users to share their multiple shoe choices with others.

Sneaker Culture

Athletic and sports footwear is now an acceptable substitute for more traditional shoes. Nike holds the largest share (19 percent) of the total footwear market in the United States (sales of $12.8 billion in 2014); in comparison, Adidas has 4 percent. Nike consumers are loyal, and they enjoy and expect technical innovation and increased performance features. Constant innovation, consumer understanding, and strong marketing images and placement are essential to maintain Nike's position. Its exclusive Flyknit technology in one-piece construction launched in 2014 and aligned with the US college football playoffs, where players wore the Vapor Untouchable football cleats that featured an ultra-lightweight Flyknit upper made with recycled polyester yarn. This linked with the target customers' lifestyle, aspirations, and sports culture.

FROM COMMODITY TO COLLECTION: THE SNEAKER AS AN EXTENSION OF SELF

Collecting as behavior originated through necessity by hoarding or saving for lean times. However, as footwear has evolved from a utilitarian product to a commodity fashion item, the practice of buying sneakers, or kicks, to create a collection is now a lifestyle activity for a core group of enthusiasts.

Devotion to collecting sneakers is a significant activity in our consumer culture today, fuelled by greater availability and rising affluence. However sneaker wearing evolved from the 1970s urban hip-hop culture in the United States, which was often a tough and poor environment where young men had to assert themselves to survive. Usually owning only one pair, cleaning them, and adding different laces for variation was common. Their rise from street style through music and sports endorsements (basketball player Michael Jordan's eponymous Air Jordans) with sophisticated communications campaigns is well documented (see the link at the end of chapter), and today's collector culture has evolved to be part of a $26 billion industry.

The increased availability of new sneakers inevitably drove up the value of original and authentic styles, leading small groups of enthusiasts to hunt out and create personal collections. Sneakers hold cultural and financial value; therefore, collecting gives people pleasure and a purpose in life and elevates their feelings of self-worth. They can buy "triples," or several pairs of the same shoe: one to wear, one to lose at home, and one to store; in many cases sneakers can remain unworn and often left in boxes. This tribe of shoppers displays both a compulsive "need" and addictive behavior.

> **" THE TROPHY SNEAKER IS THE LATEST TREND AND A PHENOMENON WHICH SEEMS TO HAVE LONGEVITY. THERE IS NOW A HUGE DEMAND FOR SNEAKERS ON THE WOMEN'S SIDE OF THE MARKET AND HARVEY NICHOLS ARE LAUNCHING A WOMEN'S SNEAKER CONCEPT, THE FIRST IN THE UK MARKET WITH A FOCUS ON FASHION SNEAKERS INSTEAD OF PURE SPORTS PRODUCTS."**
> JASMIN SANYA, FOOTWEAR BUYER AT HARVEY NICHOLS, UK

1.10

Sneakers symbolize identity and allow the person control in their sneaker "world." Collectors may often talk to the objects, favor them over family members and social situations, and find comfort in them, going to great lengths to preserve and protect them.

The activity of collection can help the transition into adulthood, which may explain why the majority of avid collectors are young men. There is a huge amount of time, effort, and commitment required to build collections, and often filling gaps—finding a sneaker previously unaffordable or unobtainable from childhood to complete the collection—is a kind of self enhancement and can help with levels of self-esteem. Collectors will queue, travel, and trade online to find exclusive products to enhance their collections.

1.10 Sneaker Festival, Laces Out, Liverpool
Many collectors attend festivals and events to meet and trade with other collectors and sneaker enthusiasts.

" TRAINERS HAVE OVERTAKEN SHOES TO BECOME THE MOST POPULAR TYPE OF FOOTWEAR PURCHASED BY MEN FOR THEMSELVES IN THE LAST 12 MONTHS . . . POPULAR AMONG A WIDE AGE RANGE, BUT MOST USED BY 45–54-YEAR-OLDS."
MINTEL UK 2015

THE PURCHASE DECISION PROCESS

The consumer goes through a series of decision-making stages when buying shoes.

1. Recognizing a "need" for new footwear, either practical or emotional (utilitarian or hedonic).
2. Identification of alternatives—can another style or a different brand satisfy the need (cultural gatekeepers)?
3. Evaluation of alternatives by considering commercial, social, and cultural influences (social reference groups).
4. Purchase of the shoes: in store, self-service, or online?
5. Post-purchase satisfaction or dissatisfaction; caring for and disposing of the shoes.

Shoppers' behavior within the store environment is important to consider, and a significant number of us shop for footwear alone. The preference between self-service footwear and service from a salesperson is roughly split in half. When trying on shoes we do not go into a changing room but try on within the store environment. In many cases the customer will look for assurance, guidance, and approval from the sales associate. If fitting is required, the relationship between seller and buyer may become quite intimate, and that has implications for retailers as it allows them to build loyal relationships with customers.

Levels of Purchase Involvement

How much effort goes into the purchase of a new pair of shoes? Much of this will depend on time, money, and the benefit new shoes will have for the consumer—especially if it has a social or reference group significance. When shopping for footwear, 74 percent of consumers rank comfort as the highest factor in their decision-making process, closely followed by fit at 67 percent. Design (49 percent), low cost (48 percent), and color (43 percent) are also key factors. Less so is the material the shoes are made from, e.g., leather or synthetic (33 percent), branding 19 (percent), and fashion (13 percent) (Mintel 2015).

As the key considerations when purchasing footwear are comfort and fit, Park and Curwen (2013) assessed the issues around fit and emotions. True fit and comfort are nearly impossible to achieve from ready-made shoes, but we learn to compromise in a way that we don't with clothing. Clothing can be altered, adjusted, and remodeled. Open sandals and flip-flops allow for variations of fit due to the nature of their design, but most footwear cannot be changed once produced. Footwear requires a high degree of involvement during purchase as it is primarily about physical comfort, performance, and aesthetics. Once purchased and worn, if the shoes are uncomfortable they cannot be altered or returned, so the risk of dissatisfaction at the post-purchase stage can be high.

ETHICS IN ACTION:
Disposal and Recycling of Shoes

Previously, footwear was an investment rather than a discretionary purchase; when shoes wore out, people took them to a cobbler to be repaired. The increased use of plastic, fast production techniques, and growing affluence have led to the slow demise of the independent shoe repair store. The Shoe Service Institute of America notes that the shoe repair industry keeps approximately 62 million pairs of shoes out of landfills every year. Stores such as Nordstrom and the Shoe Heaven floor in Harrods offer a specialist branded shoe repair service for high-value purchases, with shoes either repaired on site or returned to the manufacture to be refurbished. There are now also a number of online shoe repair services across the United States. Despite this, footwear is viewed by many as a commodity, throw-away item, and environmental issues need to be addressed. What happens to our shoes when we have finished with them? Many end up in a landfill, but there are some recycling, lateral cycling, and social initiatives that are tackling this problem.

Lost and discarded flip-flops from the Indian Ocean that wash up on the beaches of Kenya are collected by Ocean Sole and recycled in Nairobi to create handcrafted sculptures and gifts. The concept for Ocean Sole originated in 1997 as a marine environmental project to address pollution on East African beaches. Local people were trained in basic recycling and craft skills, enabling them to start making a living creating flip-flop products that are now sold worldwide.

Donating to charity or selling through eBay, vintage shops, and garage or car boot sales is a form of lateral cycling and gets rid of our low-sentimental-value shoes. Dealers buy second-hand single and pairs of shoes from thrift stores, charities, and traders who collect our unwanted footwear. Prevalent in the United States, companies like Samiyatex have several well-established routes that our discarded footwear takes, but for the most part it ends up graded and bagged, sold by weight, and shipped by container to South America, Africa, and the Middle East for resale in the second-hand goods market.

1.11

1.11 Ocean Sole recycling discarded flip-flops
Ocean Sole's process consists of collecting, recycling, and creating new products from discarded flip-flops found in the Indian Ocean.

1.12

1.13

1.14

When shoes have a high sentimental value we often keep hold of them and hoard them in boxes, garages, and wardrobes; wedding shoes and baby shoes, for example, carry emotional attachments and memories. We go through a process, known as divestment rituals, of detaching ourselves from their meaning for some time before parting with these shoes.

1.12 Clothing and footwear recycling bins
Old clothing and footwear are collected through recycling bins at local supermarkets.

1.13 & 1.14 Shoe art in Copenhagen 2014
The Wall of Shoes, an art installation by Jacob Amsgaard, was a creative endeavor as well as a way to help others. The shoes were collected over an eight-hour period, and all excess shoes and donations were given to the charity GRACE KBH.

> " FOOTWEAR IS AN EXTENSION OF SELF AND IT ALSO ACTS AS A REPOSITORY OF MEMORY AND MEANING IN OUR LIVES."
> BELK, 2003

RESEARCHING EMERGING CONSUMER BEHAVIOR

To develop a deep understanding of the consumer, a variety of sources such as market intelligence reports, academic research papers, and primary research methods are used to collect data. Retailers may use this research to develop a profile of the consumer (also known as a pen portrait) that will underpin sourcing and buying, brand development, and the marketing communications strategy. Below are three examples of how and where to find information that supports a consumer profile.

The Market Intelligence Report

Produced by market research companies such as Mintel, Keynote, and Barnes, these reports give specific, up-to-date information about commercial activities and consumer behavior happening in the market that a company operates in.

FACTORS AFFECTING THE US CONSUMER IN 2016 (SUMMARIZED FROM THE 2016 BARNES REPORTS FORECASTS: FOOTWEAR MANUFACTURING)

The United States saw improvements to its economy in 2016; however, relatively high unemployment, continuing government debt, and quantitative easing (QE) policies will keep this growth slow.

The housing market collapsed in 2008, and although it is recovering, many homes have not returned to their 2007 value. Linked to this, the wealth of the average American is at a historical low, so although interest rates are also low, many cannot afford to buy a new home. Home ownership is below 65 percent—the lowest in twenty years. The inflation rate is also low for consumer goods, but basic utilities and energy costs are well above inflation—around 11 percent—so this now commands a much larger percentage of consumers' disposable income.

US government debt and spending is high with a continued policy of QE, which in effect prints money to flow into the economy, giving a false sense of security. Currently 46 million Americans are receiving food stamps, an increase from 32 million in 2009, at the start of the recession, and 51 percent of workers earn less than $30,000 a year. Poverty levels are increasing as household income has decreased by an average of $4,000 per year. Between 2009 and 2013, student debt rose from $440 billion to $1 trillion, and 18 percent more people are collecting federal disability insurance.

Full-time, secure employment is also seen as a major problem; seven out of eight jobs created in the United States since the great recession's recovery have been part-time jobs. This trend is continuing, and many small businesses will avoid providing health insurance (Obamacare policy) by retaining fewer than fifty employees who work less than thirty hours per week.

The United States has a huge trade deficit; it imports much more than it exports, which ultimately means that very few products (with the exception of food products) made in the United States can be sold abroad to make money.

The impact of the economic environment on footwear sales in the United States is significant:
- Poverty and unemployment mean that while there is still a need for value shoes, they have to be affordable. This forces retailers to maintain low prices and source overseas, exacerbating the trade deficit.
- Consumers now have less disposable income but maintain a desire and compulsion to shop. The United States (and the West in general) thrives on consumption and material culture, and this feeds the cycle of debt and prolongs the process of QE.
- Without a thriving manufacturing base, the US economy relies on retail and services; people are still actively encouraged to spend money, despite the fact that they have little savings. Investing in long-term goals like property seems unobtainable.

Note: This is a study of the United States; there are many similar economic pressures in Europe and developed countries that no longer have a large manufacturing capacity, which will be referred to in further chapters.

Academic Research Papers

Produced by academic university research staff who operate independently from commercial companies, many of the studies are conducted over a longer period of time and are peer reviewed for accuracy and validity. Their aim is to develop new lines of inquiry and understanding, particularly within consumer behavior and retail strategy theory.

1.15 Low-priced footwear in store
Fast fashion retailers such as Primark offer a huge variety of low-price footwear, satisfying demand from both fashion enthusiasts and low income shoppers.

BARGAIN HUNTING FASHION SHOES IN MALAYSIA

Yoon Kin Tong et al. (2010) assessed how sales promotions affected motivation to purchase shoes. Malaysia is a multi-racial, tropical country not subject to extreme seasonal variations of cold and heat. Its retail sales promotions run in line with various ethnic holidays and festivals such as Chinese New Year as opposed to discounting old summer or winter stock when the weather changes, as is traditional in more temperate climates. Key findings from the study were that early visitors to the stores were often alone and strategically looking for bargains. Sale shoppers are price sensitive, regardless of their income or ethnicity; they seek out a good deal rather than indulge in the festival celebration. The pursuit of a good shoe sale with plenty of bargains will transcend religious holidays, festivals, and ethnic celebrations.

1.15

1.16

1.16 Fast fashion footwear at Matalan
Heavily discounted fashion styles on sale in
Matalan, one of the UK's leading fast fashion
retailers.

Primary Market Research

Primary market research is conducted by both commercial
and academic researchers to collect data for the reports
and papers mentioned previously. However it also focuses
on developing a direct link to understand and learn about
the potential customer. It is information or data that
has come directly from the customer and has not been
interpreted by another person or company.

Primary research is both:
• Qualitative: It gives detailed information that might
  explain motivations, values, and psychographic
  behavior. It usually involves talking to a small number of
  people in depth, often asking why, and is used to better
  understand specific issues.
• Quantitative: It gives numerical demographic data such
  as age, gender, and geographic location. It can be
  collected through online surveys or Google Analytics
  through transactional websites that can answer
  questions around who, when, and where—such as who
  is buying (male/female) and what time of day they shop
  online.

Both types of data collection will elicit demographic and
psychographic information that can be used to build the
profile of the target customer.

" QUALITATIVE TRUMPS QUANTITATIVE BECAUSE THERE ARE SO MANY
WAYS NOW, VIA BIG DATA, OF GETTING QUANTITATIVE, BUT THERE
ARE ONLY A FEW WAYS TO GET THE SENTIMENT AND FEELINGS OF
A PERSON. I BELIEVE THAT THE MARKET AND MARKETERS HAVE
CHANGED TO TRYING TO GET UNDER THE SKIN OF PEOPLE THEY WANT
TO CONSUME THEIR PRODUCTS AND SERVICES. QUANT CAN'T DO THAT,
BUT QUALITATIVE CAN REALLY GIVE YOU PEOPLE'S EMOTIONS AND THE
ANSWERS TO THE BIG QUESTION 'WHY?'"
JASON FULTON, FOUNDER, THIS MEMENTO

Developing a Pen Portrait of the Customer

By creating a consumer pen portrait, retailers can build a picture of who they are aiming their brand at, and all employees will use this as a focal point when developing the range, prices, and marketing activities. The example below identifies a typical high street shopper in her mid-thirties who buys a range of footwear for different reasons. The retailer will look to develop a range that could attract this type of customer.

1.17

1.17 Consumer pen portrait
A consumer pen portrait allows retailers to gain an understanding of who they are aiming their brand at.

Name	Jane
Age	35
Educational status	University degree followed by a master's degree in teaching
Family life cycle	Single and looking to settle down
Occupation	Secondary school teacher
Income	£35,000 pa
Where they live, with whom, and type of property	Suburb of a major UK city; lives alone in a two-bedroom apartment
Homeowner or renter	Owner with mortgage
Car access/ownership/type	Owns a vintage Mini Cooper
Personality	Fun; enjoys socializing and going on dates
What they wear and when	Modest, smart, and comfortable clothes with flat shoes for work during the day at school. Socializing on weekends: middle to high heels to look attractive with some support and comfort.
What they do, where they go, whom with socially for hobbies or relaxation	Likes watching TV with the occasional visit to the gym; enjoys a spa day and afternoon tea. Previously part of a large social group but many are now getting married and starting families.
Holidays	One main holiday per year, plus a couple of girls' weekends away.
TV/film viewing/music preferences	Netflix and new cinema releases—especially good for going on dates. Also goes to one or two summer festivals with friends.
Magazines/newspapers/radio/blogs/social media	*Sunday Times* or *Observer* at the weekend; listens to the local commercial radio on way to work; keeping in touch with friends and family via Facebook
Where they shop (which will help to identify the competition)	Main high-street stores, however finds many are too "young" and she wants clothes and shoes that fit her well and flatter her

THE ROLE OF TREND IN FASHION FOOTWEAR

Macro trends in society are identified by studying broader changes in attitude and behavior and can involve using all the senses to be hyper-alert and observant of changes, scientific breakthroughs, new product developments, and political actions. While these may not seem pertinent to shoe designers and marketers, they have an effect on consumers' acceptance and purchases, as well as a wider social impact. For example, the development and availability of the Internet, the growth of online shopping, and social media have created a platform for people to trade and resell collector's items such as sneakers on eBay. Companies such as trendwatching.com identify creative local areas, street styles, and people's behavior around the world and collate these ideas to look for common themes in business and consumer culture. As shared ideas, behaviors, and patterns emerge, they are tracked, and if the momentum carries on they are identified as trends. These trends are then adopted by people within a social system.

> " A TREND CAN BE EMOTIONAL, INTELLECTUAL, OR EVEN SPIRITUAL. AT ITS MOST BASIC, A TREND CAN BE DEFINED AS THE DIRECTION IN WHICH SOMETHING (AND THAT SOMETHING COULD BE ANYTHING) TENDS TO MOVE AND WHICH HAS A CONSEQUENTIAL IMPACT ON THE CULTURE, SOCIETY, OR BUSINESS SECTOR THROUGH WHICH IT MOVES."
> RAYMOND, 2010

Fashion Adoption

Rogers' theory, as shown in Figure 1.18, outlines how new ideas and innovation are dispersed and the speed at which they are adopted by society. Trend and market studies use this principle to identify when and how many new fashions or "innovations" are adopted by consumers. The majority of consumers (approximately 70 percent) will wait for a style to be more widely available and acceptable before making a purchase. The case study at the end of the chapter assesses the adoption of Birkenstocks as a fashion footwear style in the UK market.

The product life cycle (PLC) also helps to identify the stage a product is at within its commercial life. How long consumers continue to buy a product will determine whether it is a fad, fashion trend, or classic style. Knowledge of this is especially important for footwear designers and buyers when range planning (for further discussion, see Chapter 3). Fads are short lived with sudden popularity, are quickly dispersed (Solomon & Rabolt 2009), and happen increasingly in footwear styles due to quick and cheap production. Continued acceptance of a fad turns products into fashion trends and ultimately classic lines that last for seasons or even years. Footwear shapes such as the Oxford, brogue, and stiletto (see the list of footwear styles in Chapter 2) are enduring classics that designers repeat every season with little or no change for several years. Once a product reaches maturity, sales will steady, and as they decline the product will eventually become obsolete.

1.18

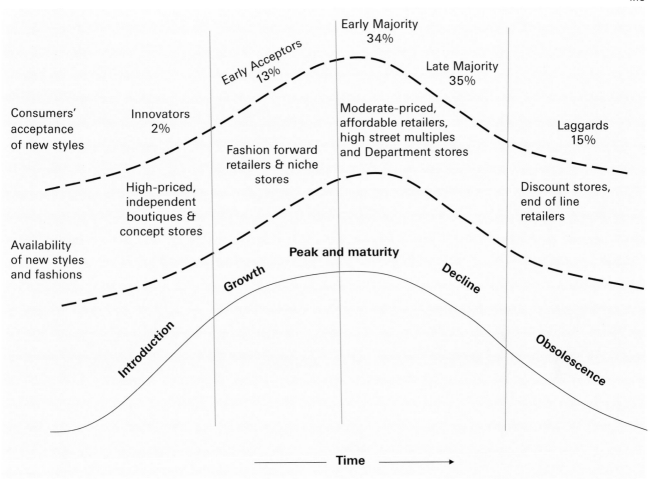

Early Majority
34%

Early Acceptors
13%

Late Majority
35%

Consumers'
acceptance
of new styles

Innovators
2%

Moderate-priced,
affordable retailers,
high street multiples
and Department stores

Laggards
15%

Fashion forward
retailers & niche
stores

High-priced,
independent
boutiques &
concept stores

Discount stores,
end of line
retailers

Availability
of new styles
and fashions

Peak and maturity

Growth

Decline

Introduction

Obsolescence

Time

1.18 Rogers' diffusion curve
Rogers' diffusion curve shows the fashion
cycle as a blend of the spirit of the times
or "zeitgeist," the relationship between the
design concept, product adoption, consumer
acceptance, and distribution.

" RATE OF ADOPTION IS THE RELATIVE SPEED
WITH WHICH AN INNOVATION IS ADOPTED
BY MEMBERS OF A SOCIAL SYSTEM. IT IS
GENERALLY MEASURED AS THE NUMBER OF
INDIVIDUALS WHO ADOPT A NEW IDEA IN A
SPECIFIED PERIOD, SUCH AS EACH YEAR."
ROGERS, 2003

Footwear Innovators and Influencers

Innovators are both the creators of a new style and select consumers who identify movements before they become known or widely acceptable. They rapidly move away from a trend once it has been adopted by a majority and constantly look for new evolutions and ideas. In many cases, innovations in footwear require technical skill, and much of the innovation still resides with the creator or designer.

The cantilevered or "heel-less" shoe, also known as the "trompe l'oeil," was originally created in 1937 by Andre Perugia. It was developed and modified by François Pinet in the 1950s and more recently seen in the collections of Antonio Berardi and Nina Ricci. Influencers and fashion

leaders such as stylists, celebrities, and fashion bloggers can be the first to pick up on new fashions like these and disperse them to a larger audience, allowing the look, style, or product to become more commercially viable. Shoes described as ugly, such as UGG boots or Birkenstocks, can elicit strong negative feelings in people, but this reaction can start trends as innovators often look to shock and re-appropriate styles in different ways.

The Flow of Fashion Trends

A fashion trend or product style may evolve in a number of ways. To gain a better understanding of how this process develops, the researcher may apply the theory of trickle down and bubble up. In many cases, fashion trends evolved from high-end designers and gradually trickled down through to the mass market and were copied and reproduced by lower-priced retailers and brands. However, designers also find inspiration from other aspects of society, particularly alternative subcultures where they would reinterpret the original use of an item that may have had particularly or exclusively functional use (see Figure 1.20).

Further examples of the bubble up theory are Punks in the 1970s wearing Dr. Martens boots, which were both functional and symbolic of the subculture they belonged to. The Dr. Martens style has bubbled up to become an accepted mainstream fashion style and brand.

1.19

1.19 Antonio Berardi heel-less shoes
An innovation in footwear, the cantilevered heel was designed to trick the eye into believing that the wearer is balancing on her tiptoes

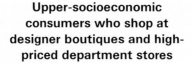

Ideas and designs from fashion innovators and leaders who create and wear unique new fashion items with distinct character

Upper-socioeconomic consumers who shop at designer boutiques and high-priced department stores

Fashion leaders and designers inspiration and creating upscale copies of anti-fashion, rebellious, or alternative footwear, e.g., Christian Louboutin studded platform stilettos

Trend-led fashion retailers who create footwear inspired and updated by high-fashion styles but is current, commercial, and appropriate for their target customer as well as being functional and durable, e.g., Kurt Geiger

Middle-level consumers who shop at mid-priced department, specialty, and chain stores

Mass fashion retailers and brands take inspiration and elements from both high- and low-fashion looks and offer a tempered version to suit a large number of consumers—pick and mix

Cheap multiple-fashion retailers that create inexpensive copies and basic styles that can be mass produced in volume, e.g., Shoe Zone

Lower-income consumers who shop at low-end chains, discounters, and inexpensive mass merchandisers

Driven by experimentation and/or necessity, new products and styles are re-appropriated, created, and worn by subcultures and street style influences, creating unusual and anti-fashion styles, e.g., biker boots, stripper shoes

1.20 Christian Louboutin studded shoes
Christian Louboutin takes influence from stripper and fetish footwear.

1.21 The trickle down/bubble up theory of fashion trends
The proliferation of the fetish or stripper shoe into mainstream acceptance over the past decade: The widening availability and acceptance of internet pornography, the sex industry as entertainment, and popular books such as *50 Shades of Grey* have made the once promiscuous or "private" footwear style acceptable. Designers such as Christian Louboutin have clearly taken inspiration when creating their footwear collections.

1.22 Stripper shoes
Once seen as underground or alternative-type footwear, the raised platform stiletto style has influenced fashion footwear designers in both high fashion and mass market.

1.20

1.22

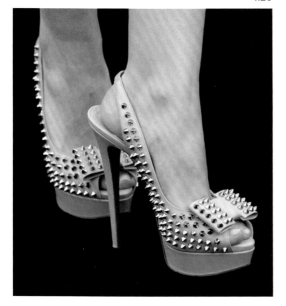

1.23 10 Corso Como in Shanghai
Established in Milan in 1990, the Italian
concept store 10 Corso Como opened in
Shanghai in 2013 as China's first international
concept store destination.

Fashion Trend Analysis

There are a number of trend prediction agencies and services that can help the designer focus on and identify some of the wider movements that are affecting fashion trends. Companies such as WGSN and Trendstop offer a fashion intelligence service that provides designers, merchandisers, and marketers a variety of trend packages and key directions of trends in the fashion and footwear sector.

Directional trend research also includes directional shopping research trips to leading, innovative fashion cities such as London, Paris, New York, and Milan as well as Tokyo, Berlin, and Copenhagen, and many others. These cities allow the design team access to the cultural movements, music, vibes, and innovations in fashion trends that are unique, and they influence design concepts as well as product ranges.

Independent fashion retailers and interiors, furniture, and concept stores such as Dover Street Market in London, L'Eclaireur and Colette in Paris, and 10 Corso Como in Milan have traditionally been destination stores for fashion trend research as they take an innovative and original approach to both store design and product selection.

Future Trends in Footwear Purchasing

The continued proliferation of new technology is allowing the customer more information and influence on how shoes are designed, produced, and promoted.

Social sharing: Allows consumers to share with a large audience behavior that would have previously been kept within their immediate reference groups. Activities demonstrated by "haul girls" and vloggers who post videos on YouTube of the products they have bought show us their emotional response to these purchases (social media is discussed further in Chapter 9).

Conscious consumption: People are increasingly aware of what they are buying, from where, and the impact it has on a community, so brands such as Toms shoes and Gandys flip-flops use their message as a fundamental reason to buy into the brand. People will always need shoes, and this gives them a reason to buy from these companies.

Collaborative consumption and customization: Customers become the designer and promoter of their own footwear. New business models such as Alive Shoes allow anyone to be a footwear creator, as does Nordstrom's partnership with Shoes of Prey, which offers a design studio in store where customers can customize individual pairs and receive them within four weeks. Elements of the shoes themselves can also be customized, such as detachable heels created by Tanya Heath Paris, which allow customers to change the height of the heel at their own convenience (see case study in Chapter 5).

> " MANOLO BLAHNIK IS AN ABSOLUTE GENIUS. HE IS NOT INTERESTED IN ANY TRENDS AND JUST DOES WHAT HE LOVES WITH SUCH CONVICTION AND PASSION; HIS SHOES REALLY ARE PIECES OF ART. CHARLOTTE OLYMPIA'S COLLECTIONS AND BUSINESS MODEL IS SO CLEVER WITH AMAZING IDEAS EVERY SEASON . . . IT'S REALLY IMPRESSIVE HOW HER COLLECTIONS CAN CATER FOR ALL THE DIFFERENT MARKETS."
> JASMIN SANYA, FOOTWEAR BUYER AT HARVEY NICHOLS, UK

CASE STUDY:
Birkenstock: Mixing Function and Fashion

Heritage of the Brand

The brand's heritage is traced back to rural Germany in 1774, where Johann Adam Birkenstock is registered as "subject and shoemaker" in the village church archives. As industrial and manufacturing processes grew in Europe, the Birkenstock family designed and developed footwear around the concept of contoured insoles with flexible arch support for better foot health. During the 1920s and '30s, affluent Europeans became increasingly aware of the health benefits of arch-supported footwear and the Birkenstock system. The company hosted training seminars and gained endorsements from leading medical specialists. Birkenstock shoes were exported to many northern and central European countries such as Denmark, Norway, France, and Switzerland. In 1947, "Birkenstock's Podiatry" was published, becoming the most widely read and published book on foot-health of its time. Firmly established as an orthopaedic footwear manufacturer, Birkenstock ventured into producing sandals in 1964 and by 1973 introduced a two-strap sandal, the Arizona, followed by a thong style in 1982. The design was created for sole support.

Establishing a Market and Spotting the Lifestyle Trend

Up to the 1970s, Birkenstock was generally unknown in the UK, and its benefits as a healthy and comfortable shoe were not recognized by consumers. During the early 1970s, a young entrepreneur, Robert Lusk, was keen to tap into the trend for functional, affordable footwear for the alternative hippie lifestyle of the late 1960s and early 1970s, epitomized by TV shows such as *The Good Life*. The "turn on, tune in, drop out" counter-culture generation were shunning the pop culture and commerciality of the 1960s and were looking for a return to nature. He identified them as the "committed brown rice brigade" and a key audience for ugly but comfortable footwear.

Lusk had a keen eye for social trends and changing consumer habits. He had trained with a Native American in the art of hand-making moccasins. In the early 1970s he traveled and spent time in Afghanistan, noticing trends for handmade and natural footwear. He had previously handmade and sold his own footwear at Kensington market and Portobello Road. Ultimately, this allowed him to open a small store in west London—The Natural Shoe Store. He sold German Birkenstock sandals in the summer and American Frye boots in the winter and introduced the Nature Shoe by Glen, a negative-heeled shoe. These were unlike anything that the UK shoe shopper had seen before; often the sandals were mocked—gaining a reputation as ugly, hippie "granola" sandals. Despite this, retailing alternative shoes was steady business and gained momentum, and in 1976 Lusk opened a store in Neal Street in London's Covent Garden, identifying it as an upcoming and affordable retail location for fashion and footwear.

1.24

1.25

1.24–1.25 Birkenstock Arizona sandal
The enduring Birkenstock Arizona sandal, designed in 1973, is now a fashion footwear classic worn by young and old around the world. It has been replicated by mass market retailers and high-fashion brands in recent years.

1.26

1.26 Alexa Chung wearing Birkenstocks in 2014
Birkenstock's fashion icon status has re-emerged and attracted the millennial generation through celebrity fans such as Alexa Chung.

CASE STUDY:
Birkenstock: Mixing Function and Fashion (continued)

Developing the Product Offer

Internationally, Birkenstock maintained its reputation for podiatry, orthopedic support, and functionality throughout the 1980s and 1990s. In the UK, its hippie associations and comfort properties did not chime with the hard-hitting power dressing and luxury brand associations of the 1980s and 1990s. Although it was still a product that was rejected by keen fashion followers, there was a core following and growing market for the sandal. In 1994, Lusk took a risk and opened a small Birkenstock shop on Neal Street. However, he needed the support of the German brand. Up to this point Birkenstock only made black, brown, and navy blue with a limited supply of white. It still saw itself as an orthopaedic footwear manufacturer, not a brand and certainly not a product that could appeal to a wider, trend-led audience. After much persuasion the company agreed in the late '90s to make two existing styles, Arizona and Madrid, available in select colors: pink, blue, green, and yellow. Lusk knew his market and the potential demand for the odd-styled sandal; the colors began selling out in the London store. Lusk's entrepreneurial talent was starting to pay off.

Becoming a Fashion Icon

While grunge fashion in the 1990s referenced Birkenstocks as an alternative style, the real acceptance as a fashion staple started in the new millennium. In 2000 there was huge interest in the sandal, and the momentum that had been building dissolved the perceptions of hippie granola sandals and re-classified the Birkenstock sandal for generation X, worn casually with jeans and military-style trousers and adopted by American film stars and celebrities. For Lusk this came as a surprise; business was good and growing, but out of the blue there were queues around the block. As the only outlet where Birkenstocks were available in the UK, demand outstripped supply and Lusk was both frustrated and disappointed he could not satisfy his customers. Also interesting for Lusk were the members of the queue for these ugly sandals that had remained the "butt" of a joke and unloved for so long by the British public. The whole social spectrum was there—city men in suits, mothers and babies, nuns, and

old and young fashion followers all rubbed shoulders together. Inevitably the media picked up on the story of the ugly but comfortable sandal and its ascendancy on to the most fashionable feet around the world. It made headlines on the six o'clock news and was featured in magazines and newspapers. This unintended exclusivity and lack of availability fuelled demand, especially as fashion editors and stylists were clamoring for a pair of their own.

In the decade that followed, the fashion frenzy production was maintained in Germany, and global distribution widened rapidly through the opening of stores. The London store was no longer the only place in the UK to buy Birkenstocks, and other retailers distributed the brand, both on- and off-line. Lusk opened the playing field; as the appointed distributor for the UK he applied the "Bazaar principle" that allowed any retailer that desired to stock the shoes, believing that the "fittest survive." Desirable fashion products and brands often control distribution by drip feeding the product into the markets to maintain demand. However with such a varied target audience across several market sectors, this strategy had little negative effect on sales. One of Birkenstock's most consistent and successful distribution points in the UK is through the shopping channel QVC.

Return of a Trend

Birkenstocks have stayed true to their original form and are now offered in 800 variations of colors and leathers, available in single, double, and multi-strap as well as toe post and clogs. In 2014 over 12 million pairs were produced for the global market. Lusk attributes this as a return to comfort and an interest in health as well as looking for dependable, good shoes that are durable and purposeful. Perceived as ethically produced, Birkenstocks are authentic, and although it is difficult to prove their orthopedic benefits, consumers are fond of them. They are a classic style and a staple for the summer wardrobe, and they are reliable and an affordable investment with very little reason, if any, to return them once bought. The British people with their quirky, irreverent style have appropriated the German haus schuh (House shoe) as a fashion style.

1.27

1.27 Birkenstock flagship store, Neal Street, London
The UK's first Birkenstock store opened in Neal Street, London in 1994

Industry Perspective:
Jason Fulton, Founder, This Memento, Amsterdam

Jason Fulton is the founder of This Memento, a global qualitative consumer research and brand strategy agency based in Amsterdam. After working for Nike for many years, Jason established the agency in 2009 and works with brands using consumer ethnographies, buddy groups, shop-alongs, ideation workshops, innovation sessions, and solution salons to develop customer insights.

What training/qualifications/experience did you have for your role?

I believe in a formal education and trusting in your gut. I have a BA (Hons) in Furniture and Product Design, retail and new business development experience, at French Connection and Diesel. My ten years at Nike gave me management experience and insights into footwear production, development, consumer culture, and innovation.

What do you think motivates people to buy footwear?

It's a quick way of showing how expressive you are to others. It's multi-purposeful and it can help expand your wardrobe because it can be so flexible with the other things you have in it.

What or who influences the consumer to buy?

Cultural influencers; real people that consumers can identify with, who've had recognizable achievements is where it's heading. The increasing visual and nonverbal nature of consumers, via their usage of social media channels, is giving the consumer access to the visual lives of celebs and local influencers. But local influencers have a closer connection to real people's lives and it's tipping towards them as being more influential right now.

Can you describe some types of consumers of the brands you work with?

Most of our consumers have been "profiled" or archetyped by a brand. They tend to be between fifteen and thirty years old and overall talented, opinionated, influential, social-media savvy, talkative, stylish, and expressive. So whether it's skate, long-boarding, football, or basketball, they're all highly informed leaders in the field of activity we select them for.

How has the consumer changed?

The consumers that we talk to are less impressed by brands: their athletes, idols, or celebrities that the brands attach themselves to. They're looking for more substance not less, more innovation and pushing the boundaries, less retro, more depth in stories and less surface. With twenty-four-hour accessibility, there's no excuse anymore for a brand to fall short of these expectations.

Which brand do you most admire and why (with reference to how they utilize consumer insights)?

I might be biased, but I'd have to say Nike. After working there for ten years and as a client of ours, we've seen their inner workings. Though they're not perfect, they do a great job of getting all of their employees fired up about who they create product, services experiences for. They articulate and circulate to everyone in their teams who the consumer is. In fact, I think they get really competitive about it. It's as if it's an incentive if a country or region can "prove" they understand the consumer more than another! If they do, that can often influence a product creation direction. *Note: Jason briefed the Nike woven shoe in 2002, which is now a collector's item.*

What advice would you give to someone studying at university and in the early stages of his or her career?

This will be one [of the hardest times], if not THE hardest time, of your career: network, network, network. Be flexible and open enough to take a diversion from the path you've chosen now and again. You might be pleasantly surprised.

1.28

1.29

1.28–1.29 Nike Air Woven HTM
While working at Nike, Jason Fulton briefed the Nike Air Woven HTM shoe in 2002, which is now a collector's item and in 2015 sold on eBay for $399.

SUMMARY

The footwear consumer is a complicated, sophisticated, and increasingly savvy shopper. Shoes carry many connotations; they are status symbols and can convey the wearer's values and lifestyle. In many ways we are judged by our footwear choice, which can evoke strong emotions. Footwear must fill numerous functions for the consumer; not only must it fit well and be durable, but it should make us feel elevated and confident as well as match outfits and the seasons.

Integral to understanding the consumer is an awareness of trends both in wider society and within fashion. Every single consumer is affected by these trends to varying degrees. How retailers and brands choose to interpret these trends is crucial to their success.

DISCUSSION QUESTIONS

1. For the consumer, can footwear be both functional and fashionable? Give examples and justify your answer.

2. What other kinds of footwear collectors are there? Are shoes more collectable because they are products that exist in their own right unlike clothes, which need the body to fulfil the intended expression?

3. Read the Birkenstock case study:
a. How has the Birkenstock consumer changed over the decades?
b. Relate the trend for Birkenstocks to Rogers' diffusion of innovation—what part of the curve is the brand at now?
c. What are the motivations to purchase Birkenstocks today?

EXERCISES

1. Discuss the evolving and changing behavior of the fashion footwear consumer, considering family life cycle and reference groups.

2. Develop a consumer pen portrait for a mid-level footwear brand using the box on page 27.

3. Go to http://www.trendwatching.com and identify the sixteen global macro trends. How do they relate to and impact on today's footwear consumer?

KEY TERMS

Conspicuous consumption—expressing one's financial wealth or status through apparel and footwear that are perceived to have a high value

Consumer pen portrait—key factors that are used to develop a profile of a target customer

Cultural gatekeepers—formal or informal methods through which the consumer is given information about products and trends to make a purchase decision

Demographics and geo-demographics—studies of a population based on factors such as age, ethnicity, sex, economic status, level of education, income, and employment. Used by brands in a situational analysis of a specific market to help inform further research and strategy

Discretionary purchase—buying products that are often for pleasure or optional

Directional shopping—trend research in fashion-forward boutiques

Divestment rituals—detaching oneself from the sentimental emotion linked to an item such as shoes before disposing of them

Emotional purchasing—buying products as a reaction to a feeling or desire

Family life cycle—the stages people go through in their lives that may have an effect on their shopping behavior and purchases

Functional purchase—buying products that fulfill a need or role

Hedonic needs—although classified a "need" or necessity, these purchases often create pleasure and emotional satisfaction

High or low sentimental value—the emotions, memories, and meanings attached to shoes after they have been worn

Influencer—someone who identifies innovative ideas and whose endorsement can affect people's decision to purchase

Innovator/Innovation—creator/creation of new ideas and trends

Motivations—the impetus to purchase shoes

One-piece construction—see Chapter 2

Per capita—by head or per person of the population (e.g., there are seven pairs of shoes sold for every person in the United States each year)

Product life cycle (PLC)—the stages a product goes through as it is introduced, grows, matures, and declines in the marketplace

Psychographics—consumer research that focuses on opinions, beliefs, and preferences to build a picture of the target group's lifestyle. Often used in conjunction with demographic-based data

Purchase decision process—the stages and choices people go through, either consciously or not, as they buy a product

Purchase involvement levels—the level of commitment and interest a person shows when buying shoes

Recession/economic crisis—financial downturn in a country's economy that affects society

Rogers' diffusion of innovation—the five stages through which ideas, trends, and products are accepted by consumers

Segmentation—creating smaller, focused sections out of a large general mass market to target a specific brand or product based on demographics and psychographics

Self-concept—a person's view about him- or herself, often very different from how others see that person

Social reference group—people, peers, family, and friends who hold great influence on products that an individual buys

Trends and trend analysis—the study of the movement of ideas, styles, and products (trends) through society and culture

Trickledown and bubbleup theory—the proliferation of ideas and designs from high-end designers to mass fashion (down) and sub- or low cultural ideas spreading upward to influence luxury fashion

Utilitarian needs—basic and practical requirements that have a functional purchase

02

FOOTWEAR DESIGN, CONSTRUCTION, AND PRODUCTION

Learning Objectives

• Introduce the footwear supply chain and explain the fundamentals of footwear design, construction, and manufacturing for luxury, mid-, and mass markets.

• Identify the global landscape of footwear production and examine the evolution of today's manufacturing locations.

• Discuss issues of corporate social responsibility in relation to labor and factory management in footwear production.

• Provide a discussion around the current and future sustainability of footwear manufacturing and new modes of construction.

2.1 Factory production at Hotter, UK
The vast majority of footwear is produced on a last; similar to the shape of the foot, it provides support for manufacturing or repair and gives the shoe structure and shape.

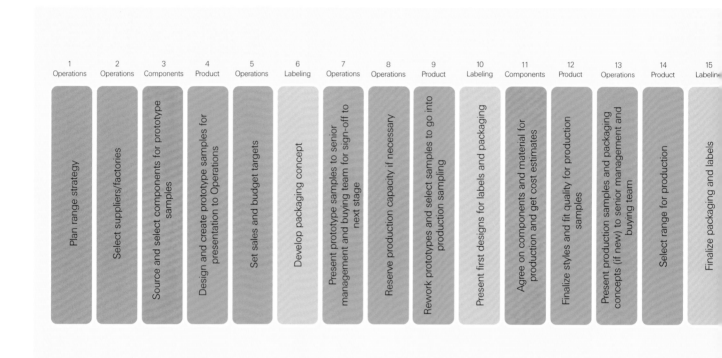

1	2	3	4	5	6	7	8	9	10	11	12	13	14	15
Operations	Operations	Components	Product	Operations	Labeling	Operations	Operations	Product	Labeling	Components	Product	Operations	Product	Labeling
Plan range strategy	Select suppliers/factories	Source and select components for prototype samples	Design and create prototype samples for presentation to Operations	Set sales and budget targets	Develop packaging concept	Present prototype samples to senior management and buying team for sign-off to next stage	Reserve production capacity if necessary	Rework prototypes and select samples to go into production sampling	Present first designs for labels and packaging	Agree on components and material for production and get cost estimates	Finalize styles and fit quality for production samples	Present production samples and packaging concepts (if new) to senior management and buying team	Select range for production	Finalize packaging and labels

INTRODUCTION

Footwear production is costly and time consuming. Historically, shoes would have been made in the country where they were destined to be sold to the end consumer, where footwear "brands" were signs of good quality and durability rather than style statements. However, in recent times the inevitable and unstoppable development of offshore production, speed of change in fashion, and consumer requirements have meant that the location of footwear production has changed dramatically.

This chapter will focus on the first part of the supply chain: the design, construction, and production of fashion footwear. It looks at the key issues around the craft-based, artisan element of footwear and how this sits with mass production, as well as new technologies that are evolving to support the sector. As the significance of branding in footwear increases, this chapter serves to put production in context when working in any part of the supply chain.

ESTABLISHING THE FOOTWEAR SUPPLY CHAIN

The first part of the supply chain involves the concept and idea planning, design, and prototyping. As a coordinated set of actions, operations, and services it comes together to deliver the final collection of footwear to the customer. The practical or tangible elements of the supply chain are concerned with production—how and what to make, location—where to make it and how to store it, and logistics—how to transport the goods and where (discussed in Chapters 3 and 4).

The information or intangible elements of the supply chain are concerned with the design, distribution, and marketing of the goods. Management of the supply chain is concerned with all of these elements and relies on the successful flow of both goods and information, as shown in Figure 2.2.

2.2

16 Operations	17 Operations	18 Components	19 Operations	20 Product	21 Labeling	22 Operations	23 Components	24 Labeling	25 Product	26 Product	27 Operations	28 Product	29 Operations	30 Operations
Decide on quantities for production based on budgets and sales targets	Agree on cost price of products with supplier	Order components for production	Agree on production plan with supplier	Place production orders	Order packing, labels and price stickers (barcodes containing 13 digit EAN code and 12 digit UPC code)	Input purchase order to retail system—order is open	Components delivered to factory for cutting and production	Packaging and labels sent to factory to be attached to goods before shipment	Production commences and completes. Quality control. Goods packaged ready for shipment	Goods shipped to retailer's warehouse	Confirm receipt of goods at retailers warehouse	Goods put out for sale	Payment of goods	Monitor sales and reorder with suppliers if necessary

2.2 The flow of goods and information from design to shop floor
Careful organization of the supply chain is also known as managing the critical path, or flow, of goods over a given time period. This figure is shown in a linear style, although some of the key processes may happen simultaneously; however, the critical points along each stream must happen in a numerical order to ensure a smooth running of the supply chain.

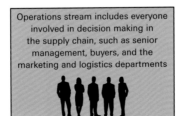

Components stream includes all raw materials that are needed to produce the end product, such as heels, leather, and insoles

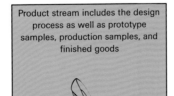

Operations stream includes everyone involved in decision making in the supply chain, such as senior management, buyers, and the marketing and logistics departments

Product stream includes the design process as well as prototype samples, production samples, and finished goods

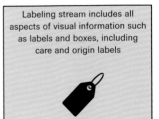

Labeling stream includes all aspects of visual information such as labels and boxes, including care and origin labels

DESIGN CONCEPTS

Creating footwear merges an artisan craft with technical engineering skill. Footwear designers and product developers have an understanding of the anatomy of the foot and walking gait in order to craft a product that is comfortable and fits properly. Whether as a high-end designer, freelance or in mass-market production they have the ability to convey their ideas in hand rendered drawings, through computer-aided design (CAD) and handcraft a prototype, pullover, or sample in 3-D prior to production. Designers must understand technical processes of construction and engineering to make footwear safe and wearable. Their knowledge of raw materials, especially leather, is crucial to create both aesthetically pleasing and durable products. They also have an acute awareness of international markets and trends in the fashion industry and find relevant and innovative ways to work alongside the apparel sector. Designers and product developers will use a variety of methods to inform their design decisions: refer to previous sales, target customers, and observe their competitors and inspirational designers as well as visit trade fairs to identify new trends and technological advancements in components and raw materials.

" MY DESIGNS ARE INFLUENCED BY ALL KINDS OF SOURCES, CULTURE, PAST, PRESENT, DECADES, ART, ARCHITECTURE—I'M ALWAYS VERY INSPIRED BY WHAT IS AROUND ME, HOWEVER I LOVE NOSTALGIA AND THE PAST—BUT TRANSLATING IT INTO MODERN."
JOANNE STOKER, DESIGNER

" IDEATION IS ABOUT QUALITY, NURTURING TIME WHERE THE CREATION OF IDEAS ARE THE END GOALS. IT SHOULD BE TWO DAYS AT LEAST, FOUR DAYS MAX, WHERE A BRAND'S TEAM, WHO HAVE PARTICIPATED IN THE RESEARCH PROCESS, GENERATE THESE IDEAS BASED ON HONEST CONSUMER INSIGHT, WHILST CONSIDERING THEIR OWN BRAND'S STRATEGIC DIRECTIONS."
JASON FULTON, THIS MEMENTO

" ALL NEW PRODUCTS MUST ALIGN WITH OUR BRAND'S CORE DESIGN VALUES. OUR COMPANY IS FOUNDED ON THE BELIEF THAT BAREFOOT-INSPIRED FOOTWEAR HELPS OUR CUSTOMERS LIVE THEIR HEALTHIEST, BEST POSSIBLE LIVES. WE USE NATURAL, SUSTAINABLY SOURCED MATERIALS THAT PROMOTE BREATHABILITY, FLEXIBILITY, AND DURABILITY. WE ALSO LOOK FOR GAPS IN OUR PRODUCT OFFERINGS AND TRY TO DESIGN NEW STYLES TO FILL THESE NICHES."
TRICIA SALCIDO, SOFT STAR SHOES

2.3 Footwear components on the factory floor
Insoles used in the production of footwear at Hotter, UK.

SOURCING RAW MATERIALS AND COMPONENTS

Component suppliers produce and supply the essential raw materials for footwear production. These encompass leather manufacturers; tanneries; fabrics suppliers for linings; and manufacturers of soles, laces, and metal fittings (nails, eyelets, etc.). Key leather producing countries are China, Italy, India, and Brazil. Trade fairs such as Premiere Vision in Paris and LINEAPELLE, Simac, and ANTEPRIMA in Milan are key events where designers source leather and footwear components as well as identify advancements in technology and fashion trends. Component trade fairs were traditionally established in close proximity to where the manufacturing of the end product would take place; however, with increased ease and speed of logistics, this is no longer critical to production. The fairs offer a wealth of qualities, colors, prices, and new innovations in leather production technology. A designer or product developer may spend many hours or even days at the exhibition working through a myriad of choices, and following this a selection of sample leathers and components will be taken back to the design office or factory where the shoe samples can be developed. Costs of raw materials can be volatile; leather prices may reflect the demand or availability of meat, and synthetic materials the cost of oil.

Tanneries

Designers also work directly with tanneries to source and produce custom-made leather. Some shoe factories in India and Brazil were originally established as tanneries that, seeing an opportunity, branched into leather garment manufacturing and footwear as markets grew.

Raw Materials

Footwear needs to be durable and purposeful, so it is made from natural materials such as leather or synthetic derivatives, or a mix of both. The use of leather and synthetic materials raises a number of environmental concerns. Although leather is a natural material that is biodegradable, cattle farming is the second-largest contributor to greenhouse gas emissions from agriculture (http://www.epa.gov). Leather is cited by many countries as being a by-product of the beef industry, therefore absenting themselves from blame, but this does not always consider the processes used by tanneries to turn raw animal hides into supple useable leathers. Curing and tanning relies on vast amounts of water and toxic chemicals such as chrome to slow down the natural process of decomposition, potentially using valuable natural resources and polluting water supplies. In recent years many tanneries have been developing systems to reduce their environmental impact.

2.4

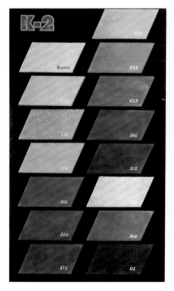
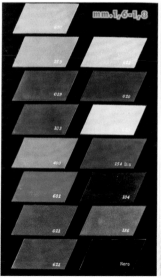

2.4 Color swatch card from an Italian leather distributor
Traditionally, Italy has been globally recognized as the highest-quality leather manufacturer due to its heritage and handcrafted techniques.

Plastic shoes fare no better. A study at MIT by Cheah et al. (2013) found that production of a single pair of sneakers creates 2.7 kilograms of CO_2 emissions, and the vast majority of the impact on the environment is incurred during the material's processing and manufacturing stages. This is the equivalent of driving a car about 10 miles, and according to timeforchange.org, in order to find a sustainable way of living we need to be producing less than 2,000 kilograms of CO_2 per year, per person.

Historically, plastics used in footwear production were not biodegradable or from recycled sources; however, recent advancements have seen a number of initiatives by both large-scale brands and independent start-ups to challenge these environmental concerns. Timberland has introduced PET plastic in their linings, uppers, and soles. It has developed a trademarked "greenrubber" from recycled sources for its soles and believes in transparency, using the responsibility section on its website http://www.timberland.com/responsibility as a source of reference.

Natural materials such as cork, obtained from the cork oak tree that does not need to be felled to harvest, are considered environmentally friendly. 01M OneMoment was inspired by the Amazon's indigenous people who painted the soles of their feet with natural latex from the Hevea tree during the rainy season; the latex would eventually wear off the foot and biodegrade. This prompted the company to design along the same principles to create a useful and functional shoe made from 100 percent biodegradable raw materials in Spain.

FOOTWEAR CONSTRUCTION

Creating a shoe is often referred to as constructing or engineering, as there are several methods of building; Figure 2.5 gives an outline of the key components needed to produce a shoe. As well as the mood boards, which highlight collection themes and inspiration, the designer must also compile a detailed spec sheet that outlines components, specific dimensions for the heel height, and construction techniques. It is not unusual to rework or refresh a classic shape from a previous season. Each seasonal collection remains contemporary, with new choices for customers, while offering the same reliable fit and quality. As in clothing, designers will work nine to twelve months ahead of when the shoes are available to the end consumer.

Retail sales staff will often have training in the construction of footwear, as much of this will have a bearing on the comfort of the shoe and is a useful sales aid, especially for shoes that are expensive or offer a unique technology aspect that can enhance their performance when worn.

2.5 Specification sheet
Each production sample has a unique specification sheet (spec sheet). This is the guide or blueprint that everyone in the design and production process, including the factory, follows to ensure that the mass manufactured product does not deviate from the original design.

2.5

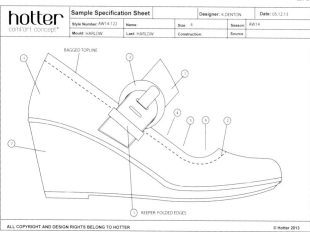

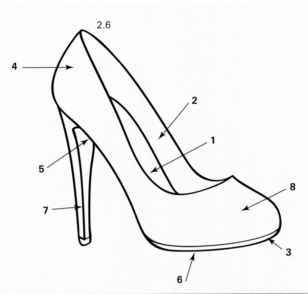

2.6

2.6 The stiletto shoe is a popular and enduring style
Footwear components used in the production of footwear.

Term	Description
1 Insole	A thin sole inside the shoe covering the outer sole and joining the upper. It is flexible, moisture absorbing, and usually made of leather or a mix with fiberboard. It should cover any nails, glue, or stitching from the construction process.
2 Lining	Covers the surface of the upper; the shoe can be half or fully lined, the choice of which may relate to cost.
3 Platform	A raised platform between the outer and inner sole made from cork, plastic, and wood; the heel component must take this into account.
4 Quarter	Two sections of the side of the shoe covering the back of the foot with a seam and reinforced by a stiffener.
5 Shank	A strip of steel, fiber, wood, or leather, inserted between the outsole and the insole at the waist to maintain the curvature of the sole from the back of the shoe to the joint and keep the heel from going under.
6 Sole	Bottom of the shoe, made from leather or manmade material that touches the ground.
7 Stiletto heel	Top–piece (heel tip that touches the ground) of 1 centimeter or less across and should have a metal reinforcement running from the top-piece to at least half the height of the heel.
8 Upper	Portion of shoe that covers the upper surface of the foot. It can be made of a variety of materials, commonly leather, plastics, or fabric.
Toe post (not shown)	Strip of leather on a sandal separating the first and second toes, holding sandal on foot either by means of a loop around the big toe or by straps to the side.
Toe cap (not shown)	Leather or fabric covering the toe or point of the shoe, usually reinforced with metal in safety footwear.
Tongue (not shown)	An extension of the vamp over the instep, below the eyelet facings. It covers what would otherwise be a gap between the facings and protects the instep from lines of pressure from the laces.
Vamp (not shown)	The front part of the upper between the toe cap and the quarters, including the toe in the case of capless styles.

The Last

One of the fundamental differences between clothing production and footwear is the use of a last. During the design phase, clothing samples will be constructed on a block or mannequin, but once production starts garments are sewn flat. To create a 3-D shape, shoes are constructed and produced on a last, a solid form usually made of wood or plastic that corresponds to the shape of the foot. It is used in the majority of footwear construction as a mold that the shoe is built around. Once the stitched flat leather uppers and sole have been attached, the last is removed.

There is a large financial investment made when developing a new shoe shape, as each pair of shoes requires a pair of lasts and each size requires a pair. One size run 3–8 UK (US sizes 6–11) including half sizes may require up to eleven pairs—twenty-two lasts. Development costs will vary, but once the style goes into production there must also be multiple lasts in each size to make quantities. The same last can be used for a number of seasons or even years, therefore reducing development and prototype costs.

Product Development and Engineering

The influence and increase of technology through CAD and computer-aided manufacture (CAM) is highly evident in the footwear sector. New entrants are expected to have a full working knowledge of CAD and its application to the design process. In essence, CAD allows the designer to view the shoes on screen in 3-D form before sampling or prototyping. CAM can be used to control elements of the automated production process, such as injecting polyurethane (PU) into the sole units, a process that could not previously be done by hand. This has increased choice, quality, and efficiency.

2.7

2.7 The last
Traditionally made from wood and carved to replicate the wearer's foot, the vast majority of lasts are now made from high-density recyclable plastic (as shown in Figure 2.1) that is made to withstand mass production and high temperatures.

OPERATIONAL PROCESSES

Although production of footwear varies depending on the design, construction, quality, and time, it is a linear process. There are several key processes that a shoe must go through once the design and prototypes have been authorized to go into production.

1. Cutting the leather or chosen material for the upper from a flat pattern. Skilled cutters know how to maximize leather hides to get the least amount of waste.
2. Stitching the pieces of material together to form the upper. The process is often outsourced to cheaper-labor countries and performed by women.
3. Sole preparation, either cutting the sole for attachment or priming the molds for polyvinyl-urethane (PVU) injection. Sole-shaped knives are used to stamp out leather sole shape.
4. Lasting stretches the softened or warmed upper and its lining onto a last (plastic or wooden foot shape), to allow for stage 5.
5. Bottoming attaches the sole to the upper through one of three methods: nailing, cementing/bonding, or stitching—or a combination of the three procedures.
6. If heels are included, this is the point where they are nailed and attached to the bottom of the shoe.
7. Finishing—this includes removing the last, polishing, trimming, inspecting, and quality control (QC); ready for stage 8.
8. Packing in tissue paper, shoe boxes, etc. which is essential to protect in transit and a key part in the branding process.

2.8

2.9

2.10

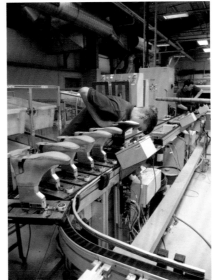

2.11

2.12

2.8 Sole preparation
Molds used for PVU injection.

2.9 Lasting
Fitting the leather uppers to the last on a moving production line.

2.10 Bottoming
Automated attachment of the PVU sole and the leather upper on a moving production line through the use of heat and injection molding.

2.11 Finishing
Removing the last from the shoes—at this stage the shoes are extremely hot.

2.12 Finishing
Polishing leather before quality checking.

Term	Description
Bespoke	Footwear that is cut and made to fit the individual wearer from his or her own unique last.
Blake sewn construction	Sole is attached to the upper and insole by a single chain-stitched seam directly through the insole inside the shoe to the outsole; primarily used in men's footwear.
Cemented construction	Sole is attached (bonded) to the upper by cement (glue) usually through heat and pressure to create a lighter, more flexible shoe.
Injection molding	Molded sole unit using thermoplastic materials such as thermoplastic rubber (TPR), poly vinyl chloride (PVC) or TPU (blend of PU and rubber); the choice will depend on cost. The process involves melting the material in the heated barrel of an injection molding machine and injecting it under pressure into the mold cavity. Creates cheap, mass-produced shoes with a flexible, waterproof, long-wearing sole.
Made-to-measure	Similar to bespoke, but can denote shoes made on a generic last that approximately corresponds to the size of the customer's foot.
Moccasin construction	A last is not needed to produce a traditional moccasin. The upper and sole are in one piece of leather, and the shoe is closed by stitching in the vamp. If worn outside, a sole and heel are attached to the bottom. However modern day boat shoes and driving shoes are either force lasted or sewn on a last and then heat set for shape.
Polyurethane	Molded sole unit from synthetic material. Expanded PU is produced by mixing two chemicals that combine inside the mold to produce polyurethane foam which is attached directly on to the lasted upper.
Vulcanized construction	Molded sole units from heated rubber pellets. Used with uppers made from suede or fabric as it can damage leather. A construction that produces a comfortable and flexible product.
Welted construction	A traditional method of producing footwear. The sole is stitched to a welt, a strip of leather running around the feather of the shoe, which is also stitched to the upper. Goodyear welted construction is complex and mainly used for high-quality men's footwear.

2.13 Construction of shoe types
There are various types of construction involved in making shoes, and the option chosen will have a direct bearing on the cost of the shoe.

Types of Construction

While a large majority of shoes are generically nailed and/or glued together, there are a number of specific techniques used to construct a shoe. Many of them have existed for centuries, such as a moccasin construction. Others evolved through the industrial revolution, such as welted footwear. More recent advancements have resulted in processes such as PU injection molding. One-piece construction using rubber and PU materials for flip-flops, clogs, and waterproof boots also gives the advantage of near zero waste in manufacturing, which lowers costs and material waste. Modern footwear construction involves heat, force, and mechanization to mass produce.

Other types of footwear include "bespoke" (or custom made) and "made to measure." Bespoke footwear is made from a last with measurements taken from an individual's foot. "Made to measure" involves using a stock last or pattern that is adjusted to fit the customer's size and design. The types of construction as listed above will have a direct bearing on the cost of the shoe. The most expensive production is Goodyear welted, and the cheaper forms are those that use man-made materials and have automated production flows, such as PU injection molding as shown in the preceding photographs.

RANGE DEVELOPMENT AND STYLES

To build a comprehensive range or seasonal collection, a selection of styles will be considered by the design, development, and buying team. A range may be built around the same or similar lasts but with different detailing on the uppers; for example, a pump and a sling back may be built on the same last and heel height but have quite a different aesthetic effect. There may also be different heel height options on the same upper, which requires a new last, but to the customer they may not look very different. Ranges consist of a selection of popular footwear styles such as flat, heels, sandals, and boots with an increasing interest in sneakers and sports-inspired footwear.

2.14

2.14 Flats
Footwear with a little (10 millimeters or less) or no visible heel.

Shoe style/term	Brand example	Description
Ballerina/ballet pump	French Sole	A closed shoe designed to resemble ballet shoe
Brogue/brogue effect	Grenson	Upper comprised of several parts, each gimped along the edges
Clog	Crocs	Originally carved out of wood but also available in variety of plastic and synthetic
Derby/gibson	Loake	Boot or shoe with the eyelet tabs stitched on the outside of the vamp for laces
Loafers	Gucci	Slip-on/no laces—a lightweight, casual shoe without fastening
Moccasins	Tods	Slip-on /no laces; similar to a loafer but not constructed on a last
Monk shoe	Cheaney	Fastened at the instep by a broad strap with buckle toward the outer side
Oxford	Barker	Closed front with the quarters stitched under the vamp
Spectators	Churches	Two-tone derby or oxford shoe
Slippers/house shoes	Mahabis	Soft indoor shoe; slip-on or moccasin construction
Espadrille	Toms	Jute rope sole with a stitched textile upper not constructed on a last

2.15

2.15 Heels
Footwear with at least 1–2 centimeter visible heel, mid-heels approximately 5 centimeters, and high heels 8 centimeters and upwards.

Shoe style/term	Brand example	Description
Wedge	Terry de Havilland	Heel is extended under the waist to the forepart, giving a flat surface in contact with the ground throughout. Often combined with a platform.
Court shoe or pump	Stuart Weitzman	Low-cut upper that covers the foot (creating a pointed or rounded toe shape) without a fastening and is stitched and/or cemented on to the sole.
D'Orsay	Christian Louboutin	A pump with sides cut away; the shoe is held on the foot by the vamp and the stiffener at the back and shows the arch of the foot.
Mule	Maison Martin Margiela	A backless shoe held on the foot by the forepart only; usually closed toe.
Sling back	Jimmy Choo	A shoe with strap passing from forepart around the hollow of the ankle, usually secured by a buckle or elastic.
T bar or Mary Janes	Manolo Blahnik	A full pump with a strap over the foot, resembling a traditional style of girls' footwear.

2.16

2.16 Sandals
Simple footwear construction in which the sole is held on the foot by an open-work upper of strips of material; can be flat or heeled.

Shoe style/term	Brand example	Description
Flip-flops/thongs	Haviana	Toe post joining two strips of material holding the foot on either side
Jellies	Melissa	Usually made from PVC using the injection molding process
Slides	Birkenstock	Leather or other material strip across the foot and attached to the sole

2.17

2.17 Boots
Footwear that covers the whole foot, ankle, and often parts of the leg; can be flat or heeled.

Shoe style/term	Brand example	Description
UGG boot	UGG Australia	Upper is cemented to the sole and usually made of sheepskin, but a cheaper version will use synthetic materials. Not constructed on a last.
Wellington boot	Hunter	Protective calf-length boot made from rubber or PVC and used in wet weather or muddy conditions.
Lace-up boot	DM	Boots with eyelets to create a secure fit and lace-up style.
Chelsea boot	RM Williams	Ankle boots with elasticated sides and fabric tab at the back of the heel to help slip on and off.
Chukka boot	Clarks (desert boot)	Low-cut ankle boot with rubber, leather, or crepe soles; two-hole lace often made with suede uppers.
Riding boot	Frye	Leather sole, pull-on knee-high boot (used for riding horses); also a generic term for fashion styles that have a resemblance.
Biker boot	Giuseppe Zanotti	Rubber sole, square toe, pull-on mid-calf boot (used for riding motor bikes); also a generic term for fashion styles that have a resemblance.
Cowboy boot	Charlotte Olympia	Cuban heel, round toe, pull-on mid-calf boot (used for workwear/farming/riding); also a generic term for fashion styles that have a resemblance.

2.18

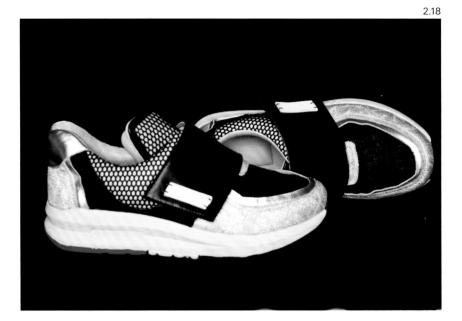

2.18 Sneakers
Sneakers were originally for sports training but the word is now a generic term for footwear made of fabric or leather upper with a leather trim in a lace-up or velcro style with rubber or cemented sole.

2.19

Adult Men and Women's Shoe Size Conversion Table M/W indicates Men's or Women's Sizes. Other systems are for either gender.																			
System		**Sizes**																	**System**
Europe		35	35½	36	37	37½	38	38½	39	40	41	42	43	44	45	46½	48½		**Europe**
Mexico							4.5	5	5.5	6	6.5	7	7.5	9	10	11	12.5		**Mexico**
Japan	M	21.5	22	22.5	23	23.5	24	24.5	25	25.5	26	26.5	27.5	28.5	29.5	30.5	31.5	M	**Japan**
	W	21	21.5	22	22.5	23	23.5	24	24.5	25	25.5	26	27	28	29	30	31	W	
U.K.	M	3	3½	4	4½	5	5½	6	6½	7	7½	8	8½	10	11	12	13½	M	**U.K.**
	W	2½	3	3½	4	4½	5	5½	6	6½	7	7½	8	9½	10½	11½	13	W	
Australia	M	3	3½	4	4½	5	5½	6	6½	7	7½	8	8½	10	11	12	13½	M	**Australia**
	W	3½	4	4½	5	5½	6	6½	7	7½	8	8½	9	10½	11½	12½	14	W	
U.S. & Canada	M	3½	4	4½	5	5½	6	6½	7	7½	8	8½	9	10½	11½	12½	14	M	**U.S. & Canada**
	W	5	5½	6	6½	7	7½	8	8½	9	9½	10	10½	12	13	14	15½	W	
Russia & Ukraine*	W	33½	34		35		36		37		38		39						W	**Russia & Ukraine***
Korea (mm.)		228	231	235	238	241	245	248	251	254	257	260	267	273	279	286	292		**Korea (mm.)**
Inches		9	9⅛	9¼	9⅜	9½	9⅝	9¾	9⅞	10	10⅛	10¼	10½	10¾	11	11¼	11½		**Inches**
Centimeters		22.8	23.1	23.5	23.8	24.1	24.5	24.8	25.1	25.4	25.7	26	26.7	27.3	27.9	28.6	29.2		**Centimeters**
Mondopoint		228	231	235	238	241	245	248	251	254	257	260	267	273	279	286	292		**Mondopoint**

Sizing and Quality

A further challenge for footwear manufacturers and retailers is the issues around size ratios. Mintel (2015) suggests that one in ten UK women are dissatisfied with shoe size options in stores. As a global manufacturing business, there is no definitive grading system (see Figure 2.19). There are huge variations, and different factories use different scales. An Asian factory might run its production in euro sizes, but the retailer may be from the United States and must label its footwear accordingly. Most brands will source from a variety of suppliers, and the customer will be unaware of which part of the range is from where but will still expect to see the same size across a range. These variances need to be addressed at the development stage and factored into production.

2.19 Shoe size chart
This shoe size chart highlights global differences in sizes. Variations in sizing can cause complications when sourcing overseas.

THE GLOBAL LANDSCAPE OF FOOTWEAR PRODUCTION

In 1996, around 10 billion pairs of shoes, boots, and sandals were produced. In 2014, global output was over 24 billion pairs. In less than twenty years, the global production of footwear has more than doubled. Traditional global footwear production patterns have evolved as the sourcing requirements of retailers and brands have changed along with the shift in global trade patterns. China is now the largest manufacturer of footwear, followed by countries such as India, Vietnam, and Brazil, which, as well as making footwear for their own growing markets, supply US and European countries with mid-/low-end men's, women's and children's footwear. Production of luxury footwear for high-end brands is still dominated by Italy. In recent years, manufacturing has been established in the emerging economies of Eastern Europe, South America, and Asia.

Ninety percent of the world's footwear is produced in ten countries, and little has changed in the past five years. The challenges faced by small and medium-size enterprises (SMEs) and large corporations are both distinct and similar. They all rely on regional, national,

and international networks for generating business and support, and (even for those whose government agenda is to increase exports) there is a constant challenge to recruit and retain staff and update in terms of machinery and technology. The three main footwear-producing continents—Asia, Europe and The Americas—are discussed next along with challenges in sustainability, sourcing, and economic impact of footwear production in their regions.

European Production

Footwear production in Europe has been in gradual decline for several decades. Mass production has been taken on by Asian factories, but it has also coincided with the growth of casual and sports footwear sales. These shoes were never before produced or consumed in the capacity they are today. Consumers' tastes and requirements have changed, and for many years established European shoe factories relied on private-label orders from retailers and wholesale distributors. As retailers reacted to the markets and changed their strategy to source from Asia, many of the traditional factories had to change to survive. They set up partnerships with factories in developing regions, and to maintain their European-based factories they developed niche techniques and brands of their own. Recession in

2.20

1. China	2. India	3. Vietnam	4. Brazil	5. Indonesia	11. Other
Pairs in millions: 15,700 Percentage: 64.6	Pairs in millions: 2,065 Percentage: 8.5	Pairs in millions: 910 Percentage: 3.7	Pairs in millions: 900 Percentage: 3.7	Pairs in millions: 724 Percentage: 3	Pairs in millions: 2,538 Percentage: 10.4
6. Pakistan	7. Turkey	8. Bangladesh	9. Mexico	10. Italy	Total
Pairs in millions: 386 Percentage: 1.6	Pairs in millions: 320 Percentage: 1.3	Pairs in millions: 315 Percentage: 1.3	Pairs in millions: 245 Percentage: 1.1	Pairs in millions: 197 Percentage: 0.8	24,300

" IN 2014 EUROPEAN FACTORIES PRODUCED 3% OF THE WORLD'S FOOTWEAR."
WORLD FOOTWEAR YEARBOOK, 2015

2.20 Footwear producers
Top ten footwear producers worldwide in 2014, by country (in millions of pairs and percentage of market share). Adapted from the World Footwear Yearbook 2015 (http://www.worldfootwear.com).

the 1990s and the global crisis in 2008 have meant that factories now focus on what they do well and leave cheap mass production to others, notably, China, Vietnam, and India. Carefully selecting suppliers that can give the best cost advantage—for example, uppers may be stitched in Poland or India and sent to factories in Italy or the UK for lasting, finishing, etc., thereby offshoring some of the supply chain—can allow for cost savings while ensuring that control is still with the brand.

Italy, Spain, and Portugal are still the largest manufacturers of high-quality footwear in Europe, producing among them 374 million pairs in 2014. Eastern European countries, such as Poland and Romania, that are more settled after political upheaval in the 1990s also have a distinct footwear manufacturing industry and serve both local and European markets with a lower-quality offer.

High-end production in European factories is still driven by demand from luxury footwear designers who use well-established factories that specialize in techniques such as heel heights and last shapes. Designers may use several factories and operate through an agent or sourcing office in Italy. These factories have a strong reputation and will often manage contracts from several designers or brands during the same production season. The Italian footwear industry employs around 80,000 people, but many Italian factories producing high-quality footwear tend to be small and family owned (often approximately ten to twenty employees). They will receive seasonal contracts from international designers and retailers and produce a few hundred pairs per week at peak season. All manufacturing operations—cut, stitching, lasting, polish and pack—are done on site. Similar to British manufacturing, this level of production is born from craftsmanship, reputation, and experience. Italy maintains a high cost per unit, with an average price of $50 per pair. This type of production is flexible but heavily reliant on networks, trust, and success of the "brand." It works for smaller designers and retailers looking for a specific technique or style on a seasonal basis but can cause supply issues for larger brands. To address this, larger companies such as Tod's own their factories, and when the Gucci group—now part of Kering—acquired the Sergio Rossi factory in Italy, it gained exclusivity and greater control of the supply chain. This trend toward greater control has increased over the last decade and ultimately leads to a vertically integrated company, discussed further in Chapter 3.

The Americas

The United States currently produces around 29 million pairs of shoes, ranking thirty-sixth in the global shoe manufacturing industry. However, as noted in Chapter 1, the country is the largest global consumer of footwear, which today presents a huge trade deficit, importing many more than it manufactures.

Much of America's footwear production was established in Massachusetts, New England, in the late eighteenth century and developed into specific factories for men's and women's footwear. Rexford (2000) notes that the initial production was of "brogan" shoes, or brogues, for plantation slaves in the West Indies and was an important part of the trade at the time. Wholesale routes were also established for businesses in the South, and shoemakers were also traders. Footwear manufacturing was established early on in the United States, and by the beginning of the 1800s, ready-to-wear shoes were available from retailers. The influence of French fashions in the middle of the nineteenth century is evident; retailers often passed American-made product as being of French origin, which was deemed to be the height of fashion at the time. It was possibly one of first instances of counterfeiting! Mechanization was imperative to the changes in production, and much of this new technology from the 1850s was used to speed up production, impacting on the production of army boots during the Civil War (1861–1868).In the latter half of the nineteenth century and early twentieth century, America was the largest global producer of footwear; much of the industry was based on the East Coast, and key areas for high-end women's shoes were around New York City, Brooklyn, and Philadelphia. Rexford (2000) notes that by the early 1920s shoe manufacturers had possibly grown too big, and the Great Depression in the early 1930s followed by World War II saw the demise of the mass production days in the United States.

" NORTH AND SOUTH AMERICA COMBINED PRODUCED 7% OF THE WORLD'S FOOTWEAR IN 2014."
WORLD FOOTWEAR YEARBOOK, 2015

Sustaining European Production:
Manufacturers' Perspective

The UK maintains a small production area around Northampton where traditional techniques are employed to produce Goodyear welted men's footwear. There are approximately 200 manual operations within the production, and it can take up to eight weeks to create a pair of shoes. Although this can be seen as artisan-style production, premium-level menswear is still produced in over twenty factories in the region.

The two biggest manufacturers of shoes in the UK are directly competing with so-called low-cost countries and succeeding. Hotter accounts for nearly one in three of all shoes made in the UK, while New Balance is reaping the rewards from making considerable investment in equipment and staff. Gina and the Florida Group have invested substantially to increase production in their UK factories to produce quality ladies footwear. The last few years have also seen new production facilities open to satisfy demand for UK-made footwear as the cost differentials and lead time required by Far East manufacturers makes European production more viable.

Production statistics, sales figures, and the media claim that footwear designed and produced in Europe is still superior quality, highly crafted, and fashionable, and there is an increasing demand from a global clientele. The reputation of a brand that is "Made in Italy" or "Made in England" is still unrivalled on a global scale, and this can be maintained by the following:

2.21

2.21 Loake
Loake has been manufacturing premium men's footwear in Northampton, UK, for over 130 years.

- Innovation—both creatively and through production techniques.
- Authenticity—being authentic and developing on the heritage of the brand.
- Market research—keeping close to the European markets and fashion intelligence but also understanding the global consumer.

" ALTHOUGH THE UK FOOTWEAR MANUFACTURING SECTOR HAS BEEN TALKED DOWN FOR MANY YEARS, THINGS HAVE STARTED TO CHANGE FOR THE BETTER. TODAY THERE ARE 5,000 PEOPLE MAKING 5 MILLION PAIRS PER YEAR, ALL WORKING FLAT OUT TO SATISFY THE INCREASING DEMAND FOR BRITISH-MADE GOODS. THE NORTHAMPTONSHIRE FACTORIES CONTINUE TO SET THE BENCHMARK FOR HIGH-GRADE MEN'S WELTED PRODUCTS. CHURCH'S, CROCKETT & JONES, BARKERS, JOHN LOBB, SANDERS & SANDERS, CHEANEYS, TRICKERS, AND DR. MARTENS, AMONGST OTHERS, ARE MAKING SOLID PROGRESS."
JOHN SAUNDERS, CEO, BRITISH FOOTWEAR ASSOCIATION

Made in the USA: Consumer Perspective

Due to the low-cost-sourcing strategy employed by many large fashion retailers and brands within the footwear sector, it is a challenge to maintain and grow manufacturing in the United States. There is a perception that goods are more expensive but better quality and support local jobs and economies. Increasing evidence in market reports also suggests that the consumer is keen to support this and is actively looking for footwear that is made in the United States. Established brands New Balance and Frye produce part of their range in the USA.

Companies such as Okabashi, based in Georgia and established in 1986, make injection-molded, recyclable flip-flops and ballet flats entirely in the United States. Making private label and developing their own brand, Oka-B has helped to spread the financial risks as well as increase ability to access different consumer segments and international markets.

Soft Star Shoes are made to order from their workshop in Oregon, and the company knows that being connected to customers is key. The need to be agile and responsive to the market is essential to survive,

2.22

2.22 Made in the USA at Oka-B
Packing and finishing at the Oka-B factory in Georgia, United States.

as flexible production with short lead times and the ability to adapt a product quickly will benefit both the manufacturer and the consumer. Soft Star Shoes' domestic production allows for total quality control; a minimized carbon footprint (from reduced shipping impacts, sustainable power sources for the shop, etc.); the creation of local, high-quality jobs and customer goodwill; and the concentration of design, production, distribution, and marketing resources all under one roof.

" **MADE IN THE USA IS DEFINITELY A PRIORITY FOR A PORTION OF OUR CUSTOMERS (ABOUT 20 PERCENT), BUT ULTIMATELY IT'S THE PRODUCT THAT MAKES A CUSTOMER LOYAL. CERTAINLY, IN THE USA, 'MADE IN USA' IS PERCEIVED AS BEING OF HIGH QUALITY. THE MAJORITY OF OUR CUSTOMERS INDIRECTLY APPRECIATE THE BENEFITS OF OUR DOMESTIC OPERATIONS THROUGH THE PERSONAL, KNOWLEDGEABLE CUSTOMER SERVICE THEY RECEIVE."**
TRICIA SALCIDO, CO-OWNER, SOFT STAR SHOES

INDUSTRIAL CLUSTERS IN BRAZIL

Brazil is the fourth-largest producer of footwear in the world, making 900 million pairs in 2014; the industry employs around 330,000 employees in over 8,000 companies. However Brazil consumes much of what it produces, and the majority of its exports go to other South American countries and the USA. Similarly to China, Brazil has the capacity to manage the entire supply chain, being able to source raw materials and components within its own country. The advantage is keeping control of the supply chain and costs down. However recent years have seen aggressive competition from Asia, and the Brazilian footwear industry is keen to protect its market. Several studies have been conducted in the Franca region, which has produced shoes for over 100 years, to try to establish a route to sustaining growth and developing a competitive advantage. There are over 1,000 companies in the region, and in 2008 they produced 28.7 million pairs of shoes. Tristão et al. (2012) studied thirty-six footwear companies operating in the supply chain: tanneries, sole manufacturers, and shoe manufacturers. Most of the sole and shoe manufacturers were relatively young, small companies, ten to twenty years old. The tanneries tended to be older, twenty to

thirty years. By their very nature, small companies are centrally managed, protective, and competitive, but the study found there is an increasing argument to suggest adopting a more collaborative and open approach to new ideas and technological innovation in regional clusters will act as a conduit for survival and success. Strategic alliances between tanneries and manufacturers need to be formed along the supply chain to stay competitive.

Asian Production

Asian footwear production for a global market has largely been driven by the sports footwear sector. As production prices rose in the 1970s and '80s in Europe and the United States, companies sought low-cost locations to make footwear. Initially, production based in South Korea and Taiwan provided US companies such as Nike and Reebok with lower costs, which gave them higher profit margins when selling product to the US market. According to Hsing (1999), during the mid-1970s–'80s, Taiwan was the largest exporter of footwear. In 1984, exports accounted for over 90 percent of Taiwan's shoe production, and over 95 percent of the shoes were destined for western countries, especially the United States.

2.23

2.23 Brazilian factory
Hand stitching loafers on the last at a factory in Brazil.

Inevitably prices rose in Taiwan, but former manufacturers were able to set up as brokers or trading companies that developed links with the new economic zones in southern China. During the latter part of the 1980s the Chinese government, under the leadership of Deng Xiao Ping, embarked on a plan to industrialize vast areas of southern China, building factories and employing migrant farmers as factory workers. This allowed Western footwear companies access to a previously untapped labor force that although unskilled was easily trained to cut and stitch as well as handle the increasingly mechanical processes of PU molding that had been developed during the 1980s. Chinese firms established partnerships with both Asian countries and eventually US and European firms that invested in machinery and training in exchange for vast pools of cheap labor.

CHINA

From the mid-1990s, China steadily grew its production capacity while maintaining low costs, giving it unprecedented economies of scale. Relationships with overseas firms were encouraged, albeit culturally challenging for both sides, which allowed Western brands to gain access to ever-cheaper products. These developing relationships were also fueled by the closing of European and US factories that simply could not compete on price and suffered due to lack of investment in machinery, time, and quality.

In 1998, China produced 4,500 million pairs of shoes, a third of the world's production. By 2014, China was producing 15.7 billion pairs per year, nearly two-thirds of global output. Vietnam quadrupled its production from 200 million in the late 1990s to around 910 million in 2014, much of this in sportswear.

During this time there was a concerted government effort to increase capacity, production, and export. Manufacturing is seen by governments as a strategy to increase the development of a country, as export sales are revenue generators and the money generated can be invested into the infrastructure of a country. Employees also pay taxes and become wealthier, giving them buying power and choice.

Traditionally, China was not a destination for fashion footwear made from leather or textiles, but with the growth of fast fashion and instant trends there has been an increasing demand from retailers. No longer able to source mid-/low-level footwear closer to home, European and US retail buyers now source directly from China, Vietnam, or India. Many premium brands such as Ugg Australia have also shifted their production to China in recent years.

" IN 2014, 88% OF THE WORLD'S FOOTWEAR WAS PRODUCED IN ASIA."
WORLD FOOTWEAR YEARBOOK, 2015

INDIA

In contrast to China, Indian footwear manufacturing has evolved in a very different pattern. As the second-largest producer of footwear today, India evolved its manufacturing from a heritage of cattle raising, leather making, and artisan crafting. Factories such as Bata were established in the 1930s by European prospectors tapping into the rich natural resources and low-cost labor, as well as the legal and political infrastructure of the British Empire. China's recent growth was contrived through strategic government planning, whereas India's evolution has largely been driven through prospective individuals and private enterprises. China's growth derived from synthetics sports footwear, which demanded high minimum order quantities and forced economies of scale. India's history as part of the British Empire, on the other hand, has instilled some of the working practices, English language, and cultural assimilations that European brands find attractive to work with. Coupled with access to the raw materials—particularly leather, knowledge, lower minimum order quantities, and low-cost labor, this makes working with India more attractive and flexible for many British companies.

New Frontiers in Manufacturing

As retail brands and buyers seek to source ever-cheaper product, there are shifts in manufacturing to less developed countries such as Bangladesh, Pakistan, and Cambodia. Parts of Africa are being funded by Chinese companies that no longer see viability in manufacturing in mainland China. Like the Taiwanese in the 1980s, new alliances and partnerships are being funded around the globe in untapped areas. Footwear production is consolidating, and suppliers who previously traded in raw materials are seeing value in producing the entire product themselves in Asia; in Europe, manufacturers are using their heritage to develop brands that could protect and maintain their businesses. In essence, strategic partnerships, collaborations, and brand building are a way of controlling and strengthening the global supply chain.

" AFRICA ACCOUNTED FOR 3% OF WORLD FOOTWEAR PRODUCTION IN 2014."
WORLD FOOTWEAR YEARBOOK, 2015

ETHICS IN ACTION:
People, Places, and Production

Corporate social responsibility (CSR) is "a company's duties and obligations to society that extends beyond ethics and legalities" (Rabolt & Miller 2009). The phrase "triple bottom line" is widely used in relation to CSR, acknowledging the importance of environmental and social concerns and financial sustainability. The role of the buyer incorporates the selection of suppliers that reflect the CSR policy of the company. Socially responsible buying/sourcing is a top priority for all companies operating in the supply chain. However with pressure to produce quick, cheap, and fashionable footwear, there is much to highlight.

There are a number of key considerations for factory owners operating in the footwear sector; many are responsible for the sourcing and cost negotiation of raw materials such as leather and plastic. Finding, investing in, and maintaining the right location, buildings, and machinery are crucial as this affects the clients in terms of sales, production, and order processing. The location of the factory and the local infrastructure also have an impact on deliveries and lead times. However, the largest variable cost is labor. Generally factory workers are on piece work, and if the factory is busy, overtime is a common practice. Production time cannot be sped up no matter how mechanized a number of processes are. Factory owners must decide where and how to cut costs.

Since 2009, China has seen substantial rising costs in traditional manufacturing clusters fueled by rising land prices, tightening environmental regulations, and increasing wages and overheads. There is a paradox for producers; if factory owners put up wages or allow unions, this creates better conditions for the workers, but prices rise. Inevitably, clients look for cheaper sources of footwear, and development starts elsewhere. This is commonly known at the "race to the bottom."

When prices in Bangladesh, Pakistan, and Indonesia go up, where do the buyers go next? Competitive retailers are driven by the pressure to be the fastest and cheapest to the market and to meet the consumers fashion demands. As soon as retailers enforce better conditions, prices go up. This leaves companies with the difficulty of having to justify price increases to shareholders and mitigating the risk of losing customers.

A move back to manufacturing in the UK or United States can ensure control, but many domestic manufacturers in Western countries are unable to compete effectively within the mainstream footwear market. These suppliers now have to offer highly differentiated products: usually either specialist footwear for specific applications or high-end designer products. Manufacturing closer to home adds value through higher quality, better fit, and greater brand value. Hotter and Okabashi, USA, take great pride in the staff they employ and provide a safe, clean, and industrious environment.

THE RACE TO THE BOTTOM IS DEFINED AS "THE SITUATION IN WHICH COMPANIES AND COUNTRIES TRY TO COMPETE WITH EACH OTHER BY CUTTING WAGES AND LIVING STANDARDS FOR WORKERS AND THE PRODUCTION OF GOODS IS MOVED TO THE PLACE WHERE THE WAGES ARE LOWEST AND THE WORKERS HAVE THE FEWEST RIGHTS."
LONGMAN BUSINESS ENGLISH DICTIONARY

" SOURCING CONTINUES TO BE AN EVER EVOLVING CHALLENGE—THERE IS A CONSTANT RACE TO THE BOTTOM WHICH DRAGS THE WHOLE RETAIL MARKET DOWN, AND THERE ARE CERTAIN CATEGORIES OF FOOTWEAR THAT WE EITHER CANNOT COMPETE ON SO DON'T EVEN TRY OR WHERE WE NEED A LOWER PRICE POINT SO WE END UP SOURCING IN INDIA, WHICH NEEDS MORE OF OUR MANAGEMENT TIME AND RESOURCES."
MANAGING DIRECTOR, BRANDED FOOTWEAR COMPANY

2.24

2.24 Factory floor at Hotter, UK
A highly automated factory floor allows Hotter to maintain production in the UK and provide a safe, clean, and supportive work environment for its staff.

2.25

2.25 Steven Stuart at Hotter
"Passion, drive, and ambition are massively important in business; recognition of staff abilities is a must. You need to get your staff to want what you want—this is crucial and the way you do this is very important." Steven Stewart, Quality Assessor at Hotter Shoes.

2.26 Chinese footwear production
Uppers being stitched in a Chinese factory.

2.26

CASE STUDY:
3-D Printed Footwear

3-D printing is officially termed additive layer manufacturing (ALM). Through the use of a coded computer program, filament cartridges, and a printer, a 3-D object is built up from the bottom layer by layer. 3-D printing is particularly appealing to industries such as footwear because it can currently be used to create prototypes, samples, and bespoke components for design concepts. 3-D printing creates a single material object that has a limited flexible function. Creating an entire wearable shoe from a single inflexible filament is still in development. The printing process is slow and therefore doesn't lend itself to mass production. One shoe can take approximately seventeen hours to print.

Traditional footwear construction uses a variety of materials in different components and considers load-bearing areas and flexibility for movement. While the technology and software are available to create shoe patterns, the materials used so far have been quite brittle. Current developments are in plastics and powder-based nylon and more recently NinjaFlex, which is a flexible, plastic-based material allowing for movement. Additive manufacturing can produce lighter-weight designs as it prints only what is needed. It also produces zero waste, only using the amount of material needed to create the product. There is also the option to grind down a 3-D print and reuse material once the style has become obsolete.

Merging traditional craftsmanship with new technology, 3-D printing allows footwear design to evolve. Iris van Herpen was inspired by 3-D printing and collaborated with Stratasys Ltd. to create twelve pairs of showpiece shoes for her 2012 collection. In 2013, Julian Hakes created the prototype Mojito shoe by using 3-D printing in the design concept phase. Hakes looked at the shoe through architects' eyes and considered the function of a shoe and then its form. The key challenge was adapting fit and flexibility but maintaining the original shape created through the 3D printer. The prototypes were worked by hand, and crafted and shaped for optimum fit and comfort—a combined effort of technology, hand, and eye. The visual aesthetic is still crucial to the success of the shoe.

The ability to create a stylish, comfort-driven, bespoke footwear offering is being explored by several start-ups. Feetz, a United States–based company, now offers the first fully 3-D printed shoe to the US market. Adidas has worked with Parley for the Oceans on a 3-D printed midsole that uses recycled ocean plastic. Although more of an awareness-raising exercise for Parley, it highlights that the capacity exists to rethink how we manufacture footwear components and shoes.

Future Trends in 3-D Printing for Footwear

- Mass customization—consumers will be able to buy patterns off the internet, download and print at home, or order from a local or mobile depot and their bespoke, personalized items will be delivered.
- 3-D printed components like customized foot beds and inner soles are now available that can combat the comfort and fit issues that traditional mass production cannot. Mobile phone apps for scanning feet can be fed into a program that could design a bespoke fit.
- The printing process will cut out some processes in the supply chain, e.g., inventory storage and logistics reducing costs and delivery times.

2.27

2.27 3-D printer
3-D printing is also known as
additive manufacturing.

2.28 Julian Hakes
The Mojito shoe collection at
London Fashion Week.

Industry Perspective:
Tricia Salcido, Co-Owner, Soft Star Shoes

Soft Star Shoes was established in 1985 and is based in Oregon, United States. The company is passionate about minimal footwear for healthy development of bones, muscles, and balance, providing handcrafted, high-quality, functional, and fun shoes. All of its shoes are made in the United States, and the company believe passionately about supporting its local environment and community.

What ethical considerations do you factor into the development process?

Sourcing materials takes into account both environmental and social impacts. We use materials that are sourced in the USA whenever possible to reduce our environmental footprint and to ensure fair and safe treatment of upstream working conditions.

What are the biggest challenges affecting footwear production in the United States today?

Obviously, for new companies, there is a large capital investment required to use conventional modern shoe-making methods, as well as astronomical differences in labor costs domestically versus abroad. Furthermore, with the exodus of manufacturing in the '60s and '70s, there remains very little local manufacturing expertise available for education and consultation. We address these issues by approaching footwear design from a novel perspective, looking for new, non-lasted manufacturing approaches that do not require massive infrastructure. We also harness the benefits of efficient direct-to-consumer distribution.

2.29

What is your opinion on the global shift of manufacturing over the past twenty years?

When we're looking at our company values and goals, we're not looking at a twenty-year timeline … we're looking far beyond. Certainly a global company can improve short-term profitability with outsourcing, but the long-term social and technological impact on its home country is likely negative. We also value "intangible" gains from domestic production, such as quality of product, quality of work environment for employees, and quality of our relationship to the environment.

For you, who holds the most power in the supply chain?

Our leather and sheepskin distributors are crucial to our operations. We are too small to source directly from tanneries, so this is a potential vulnerability in our supply chain.

2.29 Soft Star boots
A collection of boots from the Phoenix range.

SUMMARY

Footwear design, development, and production is a highly sophisticated manufacturing process with many components, complex parts, and operations, as well as varying levels of automated production. It spans the heritage craft of shoemaking to the leading technological innovations in 3-D printing.

The previous thirty years have seen a massive shift in global locations for manufacturing, with this primarily moving to China, and the sector has been heavily impacted by all aspects of globalization. We cannot return to the days where local producers can meet the voracious demands of the fashion consumer. Key considerations today for everyone in the supply chain are the availability and cost of raw materials such as leather and plastic, but most importantly human labor, which is the most integral resource. As retail margins are pushed there is an inevitable squeeze on the producers, and we must understand the ultimate consequences of this and its effect on people.

DISCUSSION QUESTIONS

1. What are the key considerations for a footwear designer/product developer?
2. How has technology impacted footwear design and production—and what does the future hold?
3. How sustainable is high-end production in Europe?
4. How can production help to drive local economies? What are the critical success factors?
5. What does the future hold for manufacturing nations in Asia?

EXERCISES

1. How does the government support domestic production in the United States? Give examples of the key manufacturing areas and show how this has impacted brands that produce in the United States.

2. How much importance do consumers place on the origin of their shoes? Consider this in relation to a) cheap fast fashion styles and b) branded footwear that has a distinct heritage.

3. Discuss from each perspective the pros and cons/issues around unfair working conditions:
   a. Consumer
   b. Retailer
   c. Retail buyer
   d. Factory owner
   e. Government/law enforcement
   f. Illegal or migrant workers

4. Go to the Loake website to watch how Goodyear welted shoes are constructed. Assess the pros and cons around the labor-intensiveness of this process. http://www.loake.co.uk/craftsmanship/

KEY TERMS

3-D printing—a computer-generated process that builds layers through additive manufacturing to create a 3-D object

Cantilevered—shoe constructed using an architectural technique (common in bridge design) of a hidden load-bearing metal shank that supports and elevates the foot, giving the illusion of no heel

Components—the key pieces, such as heels, soles, buckles, and eyelets, that are used to construct a shoe

Computer-aided design (CAD)—computer program used by designers and product developers to aid the design process

Corporate social responsibility (CSR)—the moral, social, and ethical considerations that a company considers when trading

Ideation—forming of ideas, concepts, and images that can instruct the design of a product or creative marketing campaign

Last—the form used to build a shoe around; usually made of wood or plastic

Lead time—the stage between the start and the end of the production and delivery process of a product

Off shore—manufacturing that is situated in a different country from where the brand/company is established or the product is designed and marketed

Prototype—a sample or model made in 3-D for testing before it goes into production; can also be referred to as a pullover

Race to the bottom—the continued sourcing of the cheapest labor around the world

Raw materials—the basic fabric or leather used to create footwear

Small or medium-size enterprise—business that trades in a relatively local or national level and usually employs fewer than 100 people

Supply chain—the flow of goods from design, production, and distribution to the end consumer

Triple bottom line—a company's consideration of three key factors—environment, social, and financial—for sustainability

Zero waste—a product that is created with no waste of raw materials through the production process

03

THE GLOBAL FOOTWEAR TRADE

Learning Objectives

- Explain globalization in the footwear supply chain with reference to import, export, and trade agreements between continents and countries.
- Evaluate the effect of globalization in the footwear trade on producers and suppliers via volume and value of goods.
- Identify the role of the footwear buyer and key accountabilities for sourcing and re-selling footwear, such as logistics, legalities, finance, and range planning.
- Outline and assess key methods of distribution through wholesale and vertical integration.

3.1 London Fashion Week (LFW) fall 2016
LFW exhibition held in Brewer Street Car Park, London, attracts international press and retail buyers to view established and emerging international fashion brands.

INTRODUCTION

Over the past twenty years, the developed world has rapidly increased its consumption of footwear. In order to fulfill Western appetites for more fashion at a cheaper price, this demand led to sourcing further afield to countries in the developing world. Footwear is not a straightforward product to make; it is labor intensive, and where there are large pools of willing workers, its production will thrive. Today's sourcing strategies coupled with the rapid growth of consumerism through the retail sector means that the large global retailers hold huge power in the supply chain.

A global company can make anywhere and sell everywhere; this is especially true for sports footwear brands such as Nike and Adidas. Currently a large proportion of branded and private-label footwear is designed in one country, produced in several, and transported across many international borders on route to the consumer. Trade between countries is motivated by several reasons, but primarily to generate wealth for entrepreneurs, nations, and governments. Government and political associations attempt to regulate and promote trade between countries that results in a variety of agreements and incentives as well as barriers to sourcing. This chapter will focus on the second part of the supply chain: logistics, distribution, and management of the process.

THE GLOBAL FLOW OF FOOTWEAR VIA CONTINENTS

Figure 3.2 shows the flow of goods from continent to continent and how this has evolved over a recent five-year period (2010–2014). Asia's maintained dominance as a manufacturer is evident as 88 percent of all footwear is produced there, but there are incremental rises in levels of consumption (3 percent) and importing (4 percent), suggesting the gradual increase in consumer buying power in Asia.

The small percentage decrease of imports to Europe and the United States post–economic crisis of 2008 reflects a decline in consumer demand. Africa has doubled its footwear imports, increased its production, and seen a rise in its consumption of footwear; however, it has not seen obvious growth in exports. Both South America and Oceania have stayed relatively unchanged.

Volume of Footwear Imports and Exports

The measurement and assessment of the flow of goods between counties is shown by volume, i.e., how many pairs of shoes are being traded (Figure 3.3) and value, i.e., how much the footwear is worth to the buyer and seller (Figure 3.4). These two measures are equally relevant, as the cost of shoes can vary greatly between manufacturers in different countries.

3.2 Global footwear trade in 2014
Global footwear production, consumption, export, and import by continent from 2010–2014. Adapted from the World Footwear Yearbook, 2015 (http://www.worldfootwear.com).

3.3 2014 top ten world exporters and importers by volume
In 2014, China exported over 12 billion pairs of shoes, which is 74 percent of the world's total exports of footwear. The United States imported just over 2 billion pairs, which is 20 percent of the world's total imports. This shows a clear difference between countries that export shoes and those that import shoes. Adapted from the World Footwear Yearbook, 2015 (http://www.worldfootwear.com).

" THE WORLD TRADE ORGANIZATION (WTO) DESCRIBES GLOBALIZATION "[AS] A MULTI-FACETED CONCEPT IN SO FAR AS IT DESCRIBES BOTH ECONOMIC PHENOMENA AND THEIR SOCIAL, POLITICAL, AND DISTRIBUTIONAL CONSEQUENCES."
WTO 2016

3.2

Global footwear trade in 2014

	% Production	% Consumption	% Export	% Import
Asia	88	52	86	26
Europe	3	17	11	38
North America	2	15	1	23
South America	5	7	1	3
Africa	3	8	1	8
Oceania	0	1	0	1

Global footwear trade in 2010

	% Production	% Consumption	% Export	% Import
Asia	87	49	85	21
Europe	4	20	11	43
North America	2	17	1	28
South America	6	8	1	2
Africa	2	5	1	4
Oceania	0	1	0	2

3.3

Exporters			Importers		
Country	Pairs in millions (volume)	World share %	Country	Pairs in millions (volume)	World share %
China	12,125	74.1	USA	2,350	20.5
Vietnam	758	4.6	United Kingdom	679	5.9
Hong Kong	241	1.5	Germany	633	5.5
Germany	228	1.4	Japan	590	5.1
Indonesia	228	1.4	France	509	4.4
Belgium	227	1.4	UAE	417	3.6
Italy	219	1.3	Italy	334	2.9
India	200	1.2	Hong Kong	323	2.8
Turkey	176	1.1	Spain	316	2.9
Netherlands	165	1	Russian Federation	311	2.7

Value of Footwear Imports and Exports

The world's largest footwear exporter by volume and value is China, whose main markets are the United States and Europe. In 2014 China exported 74 percent of the world's shoes; however, this is only 40 percent of the total value of the footwear trade, suggesting that many of the shoes exported are low-value items. Increased labor costs means that buyers have looked to other developing nations such as Vietnam and Indonesia to source low-cost products. With a relatively good infrastructure and trade agreements in discussion, many of the more established and stable Asian countries like Vietnam are well placed to pick up the export relationships that China may not be able to maintain due to rising costs.

European countries are responsible for only 3 percent of global production, but Italy, Germany, and France are important global exporters by value with many high-end and luxury brands made, finished, and marketed in Europe to a global audience. Import and re-export trade is also significant in countries such as Belgium and the Netherlands, which export to other European countries in the region. Italy exports 1.3 percent of the world share of

3.4 2014 top fifteen world exporters and importers by value
In 2014, China exported over $5 billion worth of footwear, which is 40 percent of the world's total footwear exports, but the value of the shoe trade also includes more European countries that are responsible for 30 percent of the total export trade—this suggests that the cost of shoes is higher when sourced from Europe. Imports are also more diverse and reflect the consumer appetite for fashion and higher-priced footwear across the globe. Adapted from the World Footwear Yearbook, 2015 (http://www.worldfootwear.com)

3.4

Exporters			Importers		
Country	$ Millions (value)	World share %	Country	$ Millions (value)	World share %
China	53,837	40.5	USA	26,595	21.8
Vietnam	12,200	9.2	Germany	10,028	8.2
Italy	11,138	8.4	France	7,437	6.1
Belgium	5,566	4.2	United Kingdom	7,110	5.8
Germany	5,166	3.9	Italy	5,504	4.5
Indonesia	4,761	3.6	Japan	5,452	4.5
Hong Kong	4,014	3	Russian Federation	4,318	3.5
Spain	3,540	2.7	Hong Kong	4,288	3.5
Netherlands	3,295	2.5	Belgium	3,676	3
France	3,095	2.3	Netherlands	3,621	3
India	2,610	2	Spain	3,175	2.6
Portugal	2,452	1.8	Canada	2,408	2
United Kingdom	2,079	1.6	China	2,031	1.7
Romania	1,374	1	UAE	2,008	1.6
Slovakia	1,226	0.9	Austria	1868	1.5

pairs but has a global value of approximately $11 billion (the third-largest in the world at 8.4 percent), indicating that shoes exported from Italy are high-value items.

Despite being the second- and fourth-largest producers of footwear, respectively, both India and Brazil have relatively low exports, with much of their low-cost product consumed within their own nations. It is difficult to compete with China as a low-cost exporter or establish as a premium manufacturer in line with European production.

TRADE TARIFFS

Trade tariffs are taxes imposed on goods and services. They are most commonly placed on imported goods and, as such, are used as a way of restricting trade due to the fact that they increase the price of imported goods and services.

" A TRADE TARIFF IS A TAX OR DUTY WHICH IS PLACED ON GOODS CROSSING POLITICAL BORDERS (OR CUSTOMS UNIONS). IMPORT TARIFFS ARE THE MOST COMMON, AND INVOLVE A TAX BEING ASSESSED ON PRODUCTS COMING IN FROM ANOTHER COUNTRY."
EUROPEAN COMMISSION, 2014

Categories of Import/Export Behaviour by Country Typical in the Footwear Sector

Over time patterns of trade evolve, and the current patterns in the footwear trade suggest that countries fall into one of three categories: those that make and export footwear and have a strong reputation and capacity to do so, either historically or with large resources; those that import and consume, typically those whose local manufacturing base has declined, meaning it is no longer cost effective to make shoes at home; and those that are traders—acting as a hub by both importing and exporting large volumes to resell in more local markets.

Make and Export

Developing countries with large populations that have used local footwear manufacturing to satisfy demand at home and create revenue through exporting include China, Vietnam, Indonesia, Portugal, India, Romania, and most recently Turkey. Italy, although well-developed, has maintained its unique position as a premium footwear maker and exporter.

Import and Consume

This category includes established countries with large populations that previously satisfied demand by producing locally; the cost of manufacturing is now prohibitive, and consumer appetite is high. There is now a dependency on low-priced footwear from overseas and a trade deficit. The United States, Germany, France, UK, Japan, Russia, Canada, South Korea, and Austria are key countries.

Trading Posts

These are established countries with small populations, mixed resources, and historic manufacturing that now act as brokers, importing low-priced footwear and re-exporting to regional markets in Europe, Belgium, the Netherlands, and Spain. Hong Kong has traditionally been a trading hub for both Europe and the Americas.

CASE STUDY:
European Union Trade Tariffs in Footwear

The European Union (EU) is the largest single market in the world, with twenty-eight countries in 2016* and around 500 million consumers. Goods bought and sold within this area are not subject to internal duties but are classified as global import/export. There is one common tariff applied to each classification of goods entering the EU.

Until 2005 there had been a global structure of quotas in place, managed through the Multi-Fibre Arrangement (MFA) and the WTO to protect established producers of clothing, apparel, and footwear in their own country or union. In January 2005, the quotas were dissolved and there was a surge of footwear imports to the EU (700 percent) from Asia. Claims of "dumping" cheap shoes and selling goods into Europe at below-cost price were levied at Chinese and Vietnamese manufacturers. The EU voted to impose a number of duties on footwear imports and introduce an anti-dumping order to protect existing footwear manufacturers operating in the EU.

Production countries such as Spain, Italy, and Poland claimed to be affected by the cheap Asian imports and in 2006 voted to support a 16.5 percent duty on goods from China and 10 percent from Vietnam for two years. The UK, with limited low-end production, voted against the duties. In 2011, after inquiry and debate, the duties were abolished and there is now no limit or additional tax on the amount of footwear imported into the European Union from any country.

3.5

3.5 Spanish shoe shops
Handmade shoes in Spain from chrome-free and recycled leather that supports local producers.

The duties were put in place to protect European manufacturing, but the volume of product imported by retailers from China and Vietnam could never be met by existing manufacturers. The cost of the duties was ultimately passed on to the consumer, and many of the non-manufacturing nations believed that it had little effect on their choice of supplier. They believed extra duties were a risk to consumer spending as the cost is ultimately added to the retail price. In some cases it has prompted the price-led retailers to source from other countries such as India, Cambodia, and Bangladesh, as the buyer is ultimately looking for the cheapest producer.

Furthermore many European brands that developed and invested in factories in China became very frustrated. Danish comfort footwear brand ECCO invested 20 million Euros in Chinese factories to produce up to 5 million pairs per year. In 2008, there were 478 footwear plants in China that had received direct investment of $737 million from European companies, most of which export back to Europe. Large, well-established brands such as Adidas, whose design and head office operations are based in Europe, are truly global operations and often export much of the product that has been imported to Europe.

In their report, Dunloff and Moore (2014) concluded that "Anti-dumping duties [quotas] are likely to harm those EU firms that are most integrated into the world economy, and provide only temporary and, at best, only limited relief to EU firms that produce exclusively for European consumers."

*In 2016 the UK voted to leave the EU (Brexit). Once the exit process has taken place it is likely that there will be an impact on trade for many countries both within and outside of the European Union single market, which is as yet unknown.

FREE TRADE AGREEMENTS

The United States places a variety of duties on leather footwear products imported from overseas, typically ranging from 6–10 percent but in certain instances as high as 37.5 percent. US buyers are keen to source volume footwear from cheaper manufacturers, so this allows for negotiation with countries that would like to cooperate to have access to one of the largest and most affluent footwear markets in the world. The United States has 20 free trade agreements (FTAs) with several countries in Central and South America and the Middle East, as well as Australia, Singapore, South Korea, and Morocco, and is currently in negotiations to establish a Trans-Pacific Partnership (TPP) with several Asian countries.

In 2006 Morocco and the United States entered in to a bilateral FTA that allowed duty-free trade of footwear between both countries under certain conditions. At least 35 percent of the product must have been made in either country and when finished must be shipped directly to the other country. The agreement must see mutual benefit and growth of trade between both nations. Over the last five years (2014 figures) Morocco's largest trading partner was the EU, with over 86 percent of its exports going to Europe; this is due its close proximity and quick lead times. Despite government interventions, fashion footwear moves fast and buyers will always seek the supplier that satisfies their customers' demands.

ETHICS IN ACTION:
The African Footwear Trade

Rising production costs in China have created an industry-wide concern for the future of its manufacturing. China needs to find alternative production to meet its clients demands and, prompted by relationships with the United States, its interest has turned to Africa, notably Ethiopia.

Ethiopia has a strong (but underdeveloped) agricultural base and is an established leather producer, known for exporting hides and skins. As a developing nation, footwear manufacturing has traditionally been done by micro- and small businesses serving a local market; however, since 1993 a new government regime reduced import duties. In a study by Gebre-Egziabher (2007), she suggested that the government believed that Chinese shoes were superior in design, price, and quality, with the result that they have taken over the domestic market. The Ethiopian footwear industry is now challenged with cheap imports and second-hand footwear cast-offs from the West (lateral cycling; see Chapter 1). Coupled with this is an increasing desire from international manufacturers to access natural resources, raw materials such as leather, and, with a population of around 80 million, a potentially cheap workforce.

The Ethiopian government has welcomed partnerships and purchases of Ethiopian-owned companies and actively supports Ethiopian entrepreneurs seeking investments from China to develop the footwear industry. American organizations such as Pace International Inc. work with the United States Agency for International Development (USAID) and international development agencies as well as footwear brands such as Browns Shoe Co. to coordinate sourcing components from China and production in Ethiopia.

3.6

In light of the United States' and China's interest in Africa, Ethiopia must consider ways in which it can take advantage of its strengths while avoiding harming its resources and people. Gebre-Egziabher (2007) suggests that local producers should collaborate with investors and government and that "training, technology, quality control, benchmarking and reorganization of production should be designed as a package of intervention."Pace also states similar concerns, using footwear production to alleviate poverty, foster self-sufficiency, and equip workers with key skills. Its mission is to create commercial investment opportunities and resources for governments, world producers, and customers that generate wealth (http://pace-inc.com/our-mission/).

3.6 African shoe manufacturer
For companies shifting production to Africa, the partnerships should be sustainable and allow wealth to stay in the country to reinvest and develop an industrial infrastructure for the greater good of the society. Workers should be paid a fair price and profits should be distributed fairly.

THE BUYER'S ROLE IN SOURCING TREND-LED FOOTWEAR

The role of the retail buyer in fashion footwear is diverse, and knowledge of fashion trends and competitors is essential. They must be able to liaise with suppliers as well as internal departments, especially important in large retail organizations. They must be able to present their selections to senior management and sales staff as well as skillfully negotiate with suppliers, which involves calculating sales and profit and costing.

The implications of making an incorrect buying decision may be greater in retail organizations than consumer markets. Therefore, footwear retail buyers proceed with caution and are less likely to make impulse purchases. Marketing and sales managers need to develop an understanding of buyer behavior and the retail structure that the buying group is in. The key challenge for sales executives is to satisfy the diverse requirements of the store buyers.

Footwear buyers generally fall into two categories, those purchasing brands for resale and those who work with in-house or freelance designers and product developers to source styles to adapt for their own private label. Large retailers—both nonspecialist and footwear retailers—as outlined in Chapter 4, develop, source, and retail their own store brands. These large buying organizations, such as supermarkets (e.g., Walmart and Tesco) and fashion multiples (e.g., Marks and Spencer and New Look) have developed considerable power over global suppliers through their economies of scale and have been able to pressure suppliers for lower prices and longer terms of payment.

Branded Footwear Buying

Branded buyers are employed by department stores, multiples, independents, and e-tailers; their role is primarily to select finished merchandise from brands. They must be able to identify emerging designers and brands and be prepared to take a gamble on an unknown brand. They scour trends everywhere—hotels, restaurants, street style, red carpet, Instagram, and even their own customers. They must know how to edit a designer's collection, select the appropriate product for their store, as well as know when to move on and edit out brands. There is a strong emphasis on brand integrity; in many ways they are a custodian of the brand on behalf of the designer. They need to find a balance between their collections, and adjacent brands must complement each other. Knowing their customers means they may offer a mix of high and low prices and broad or shallow range, depending on their objective. They often take part in educating and empowering the customer to make them feel confident in what they have chosen (see Chapter 1 Industry Perspective).

3.7

3.7 Valentino at Harvey Nichols
Red camouflage Valentino flats with matching clutch bag, bought as part of the fall 2015 shoe range in Harvey Nichols.

Advantages	**Disadvantages**
Few design, product development, or logistics issues	New brands may not materialize six months later due to production and delivery issues
Reduction in stores' own marketing costs (visual merchandising, PR, creation of advertisements)	Lack of historic sell through, so reliant on brand strength rather than product in the beginning
Trade leverage in bargaining with other brands	Branded supplier can demand a set budget and costs can rise
Higher ticket price/recommended retail price (RRP)/manufacturer's suggested retail price (MSRP)/margin	Over-dependency on one brand for delivery, which can be problematic if they are late, have poor design and distribution, or open up more stockists or go online
Kudos	

LOGISTICS AND LEGALITIES

Importing and exporting footwear across international borders requires adherence to a number of key documents. These documents must be presented during the shipping process and clearly show the origin of goods and transaction between buyer and seller to allow for a straightforward delivery.

- Commercial invoice, which is proof of the transaction between the two parties and is similar to a sales invoice—details the product, quantity, and value of the transaction.
- Freight documents, usually a bill of lading or an air waybill, which outlines who has responsibility for the goods in transit and when.
- Freight insurance, which covers situations such as damage during handling, storing, loading or transporting cargo, as well as less common risks such as riots, strikes, or terrorism.
- Packing list, which is an inventory record of all the goods in the shipment.
- Customs value declaration (EU) is needed if the value of the goods exceeds 10,000 euros.
- Single administrative document (EU) details the goods and their movement around the world, essential for trade outside the EU, or of non-EU goods.
- Commodity Codes (EU), also known as Harmonized Tariff Schedule (HTS) codes (United States) are used to classify goods to work out the amount of duty that should be charged.
- CITES (the Convention on International Trade in Endangered Species of Wild Fauna and Flora) documentation, outlining the origin of the raw materials used and whether it is legal to trade. Some snakeskin, such as python, is not allowed into California.

Labelling Footwear

It is a legal requirement that products be shipped and re-sold in European stores with a label attached to at least one of the shoes to indicate the material it is made from. If using symbols, the retailer must display the explanations in store. The manufacturer is responsible for applying the label before shipment; however the retailer is responsible for ensuring it is selling what the label says. Origin labeling of footwear, although not a requirement in the EU, is a legal requirement in the United States.

3.8

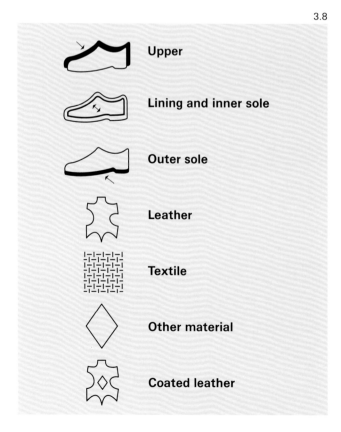

3.8 Labeling footwear
All footwear imported and traded within the EU must detail what the shoes are made from using the symbols shown here.

Incoterm	Named place	Sharing of costs and risk between buyer and seller in international traffic
EXW Ex Works	Loading location	
FCA Free Carrier	Loading location	
FAS Free Alongside Ship	Port of loading	
FOB Free On Board	Port of loading	
CFR Cost and Freight	Port of destination	
CIF Cost Insurance and Freight	Port of destination	
CIP Carriage and Insurance Paid	Delivery location	
DDU Delivery Duty Unpaid **DDP** Delivery Duty Paid	Delivery location	

3.9

Seller's cost/risk   Buyer's cost/risk

Incoterms

Goods can be shipped by road, rail, air, train, or sea and are usually shipped using a combination of several of these. Incoterms (International Commercial Terms) are an agreed-upon set of instructions detailing who is responsible for each stage of shipment, costs incurred, and delivery details. Most common terms in footwear would be Ex Works (EXW), the buyer is responsible for the goods once they have left the factory; Free On Board (FOB), the seller is responsible to deliver the goods to the main form of transport whereafter the buyer assumes responsibility through customs; Delivered Duty Unpaid (DDU), the seller is responsible for the entire delivery and the buyer is responsible for any duties incurred.

3.9 Incoterms 2012
All goods that are moved across geographic borders must use one of these three-letter codes to outline how the goods are traded and shipped between buyer and seller.

THE COST OF SHOES

For the retail buyer, "price" is associated with cost of goods and turning them into a commercial product, maintaining cash flow, and making a profit. However, the customer sees "price" from a different perspective. A product or a brand has both emotional and functional benefits, and price serves on a number of levels, and so it is influenced by demand.

There are three fundamental questions when selecting product to retail: How much does it cost to get the product to the shop floor? What is the gross margin? How much is the customer prepared to pay?

When setting prices, a company must consider its overall objectives in both the short and long term. Do the costs of materials, distribution, promotion, etc. fit with the ability to make a commercially viable product? How can the cost of goods maximize sales and cash flow as well as improve competition in the market by matching, cutting, or increasing the prices? Considerations will also include taxation and legislation such as VAT (Value Added TAX) / sales tax. Is the price consistent with image and can the product withstand price promotions such as discounting without devaluing the brand image? For many luxury brands, service and exclusivity may offset price, and the price set may not relate directly to the cost of the goods. This chapter will cover the costs associated with sourcing footwear, with further discussion on pricing strategy for retailers in Chapter 5.

Setting Budgets for Buying Shoes

Budgets are determined according to the number and performance of each retail outlet. Shops with the highest turnover usually receive the largest budget allocation. There is a minimum stock required to break even per store or department. Budget plans may also consider seasonal promotions, peak selling times, and discounting unsold stock at the end of the season. Buyers must also be aware of current stock holding and whether anything can be carried over to the next season and the sales targets the company is aiming to achieve. How much stock is on hand and how quickly can it be reordered? "Open to buy" is the term used for the money available to buy new or reorder products within a current season that is not stock already received or committed to in a forward order. Buyers must consider what stock they already have, as money can become very quickly tied up in unsold stock, which will affect cash flow.

Costings and Calculations

Cost price (CP) is cost of goods bought for resale; this may or may not include associated costs such as delivery, duties, and insurance.

Selling price (SP) is the price the goods are sold at in the retail market; this may be determined by what the market can take and will be adjusted to consider competitors and customers.

Selling price (SP) = cost price + gross margin (GM)

Gross margin (GM) is the percentage of the selling price that represents store running costs, wages, shipments, and net profit (NP).

Markup (MU) is doubling or tripling of the cost price to get the selling price. Most footwear brands will recommend a selling price using a 2.5 or 3 markup. This ensures there is room for the retailers to cover costs and meet their gross margins but also gives parity across a number of distribution outlets. The higher the markup, the larger gross profit.

Markup examples

$$CP \times 2.5 = SP$$

CP \$20 (£15) X 2.5 = SP \$50 (£37.50)
CP \$20 (£15) X 2.7 = SP \$54 (£40.50)
CP \$20 (£15) X 3 = SP \$60 (£45.00)

55% GM = CP/0.45 = SP
e.g., \$16 (£12)/0.45 = \$35.55 (£26.60) SP

A 50 percent margin means it is a 100 percent markup (from the cost price); therefore a 50 percent margin is higher than a 50 percent markup, which is only adding on half of the cost price.

Many retailers will require a minimum margin percentage in order to cover all their costs and still make a profit; they will require between a 2.5 and 3 markup for new brands and depending on market level.
• Note:
  A lower markup of 1.88 for sports trade and 2 for fast fashion shoes may be used as they are sold in higher volumes than designer footwear.
• Prices may be rounded up or down to the nearest whole dollar when retailed to the customer.

Markup examples

GM % = (SP − CP) x 100/SP
(\$66 (£50) − \$30 (£22.50) = \$36 (£27.50))
\$36 (£27.50) X 100 = 2,750/\$66 (£50) = 55% GM

60% GM
e.g., \$59 (£45)/0.4 = \$147.50 (£112.50) SP

Worked examples:

Made-in-England luxury women's footwear brand wholesaling to Europe

- CP Outsource production to a factory; cost price per pair = £150
- MU Markup by designer/brand (20%–30%) = £45
  - (Includes all design-related research, travel, marketing, trade fairs, and salaries. After these have been paid, what is left is net profit.)
- WP Wholesale price to sell to retail buyers, boutiques, departments stores, etc. = £195
- MU Markup required in store is approximately 2.5
- SP is £490
  - (includes all shipping, buying-related costs, travel, wages, and all store running costs, rent, salaries, marketing and VAT/sales tax. After these have been paid, what is left is net profit.)

Fast fashion clothing retailer selling PVU flip-flops sourced from China

- Purchased directly from a factory in China; cost price per pair = 20 RMB (Chinese Yuan)
- Cost to land goods in the EU (shipping and import duties) per pair = 4RMB, including exchange rate
- CP Cost price of goods to arrive in warehouse = $2.50
- Markup required in store is approximately 2
- SP is $5

Factory-owned brand in Italy and exporting internationally

- Cost of production, wages, factory operations, sales, and marketing per pair = 60 euros
- Ex-factory price is 60 euros per pair, and distributors and retailers are be expected to cover all shipping, duties, and taxes.

Invoices, Discounts, Payment and Credit Terms

Once production is agreed on, order confirmations are raised for both parties outlining the terms of sale. These sales orders or order confirmations generate the invoice and include relevant information for shipping, delivery and payment—name; address; date of issue; invoice number; and description (name, quality, etc.), amount, and value of the goods. Shipping details, incoterm, and transport are also included. The invoice includes value and currency of payment and the terms of payment (method and date of payment and discounts).

Discounts from suppliers are negotiated by the buyer and can be either:

- A percentage off the cost of goods based on size of order—for example, 5 percent off the cost price for orders over 10,000 units.
- A percentage off the total invoice, known as an early payment incentive—for example, 5 percent off the cost of the invoice if paid in full within ten days of receipt.

Larger store buyers may prefer to see the percent off cost of goods, as this will increase their profit margin; if it is taken off the invoice it may be included in the savings made by the company overall. Independent retailers are incentivized by early payment as they buy smaller quantities.

" THERE IS NOW A MASSIVE PRESSURE ON MARGIN. OUR WHOLESALE CUSTOMERS (RETAILERS) REQUIRE MORE MARGIN TO CATER FOR THE CHANGING MARKET, E.G., VOUCHER CODES AND PRICE DEFLATION. THEY ALSO NEED TO MAKE RETURNS FOR PRIVATE EQUITY INVESTMENT, SO THEY PASS THAT ON TO THE NEXT IN THE SUPPLY CHAIN, I.E., BRANDS LIKE US. WE THEN IN THEORY PASS ON TO THE FACTORIES, BUT YOU CANNOT PASS IT ALL ON WITHOUT ENDANGERING QUALITY OR SERVICE."
MANAGING DIRECTOR, BRANDED FOOTWEAR COMPANY

Invoices for payment can be presented as a pro-forma if no credit exists with new or small retailers, and payment for goods is required in advance or as the goods are ready. Invoices are raised for retailers with credit terms once the goods are shipped. Credit may be extended for upwards of thirty days, and it is not unusual for large, powerful buying groups to negotiate up to 120 days (three to four months) before paying for goods, demonstrating the power that retailers have established over their suppliers.

3.10

Monitoring Retail Sales

A key planning tool for buyers is to monitor and review the weekly "sell through report." The report will be generic to each retailer and collates sales figures through electronic point of sale (EPOS) or till points in store. Known as WSSI (weekly sales stock and intake plans), it shows stock location, both delivered and bought or committed to be delivered (forward ordered). It highlights amount of stock "on hand," which is the average rate of sale per unit worked out in how many weeks' cover available, i.e., if the shoes keep selling at the same rate, how many weeks of stock is left before a reorder is made or new stock is delivered. These figures help to manage the day-to-day sales and stock in the business for merchandisers and help make decisions surrounding future product selection, reorder, and markdowns.

Range Planning and Selection

Once budgets have been set, buyers will start to select product and develop ranges for the forthcoming season. Comparative shopping is an audit of competitors' products, pricing, and retail environment. It focuses on a comparison between similar products and ranges within the stores and allows buyers to gauge where their products sit in the market. It may also highlight opportunities that have been missed in the competitors' offer that a company could fill. Directional shopping, as highlighted in Chapter 1, allows the buyers to identify and forecast future trends.

Selection of products will differ depending on the reason for ordering the products: reorder or stock fill in, new season product from existing supplier or brand, or new brand. Selection may also depend on size of the shop or area in the store where the product will be on display. There must also be a mix or balance between classic and fashion styles, including seasonal variations and a proportion of different types—flats, heels, boots, sandals, etc.—and consideration of size ratios and allocations.

3.10 Factory invoice
Invoice from an Italian shoe manufacturer to a customer in the United States, detailing terms of sale; EXW Ex Works and payment terms of sixty days credit. The invoice also includes style, color, quantity of shoes, and price in euros.

Figures as of: Saturday, 7/26/2008

Sunday, July 27, 2008 4:46 PM

Dollars in Thousands

Weekly Selling - By Multi Virtual Style - Vendor Style
Office ID: B0770819 - SP08

Category: Regular,Markdown,Clearance,Final

Request name: 120894 - SP08

	Retail Amt	Cost Amt	Rec'd 1st/Lst	Sls TW	Sls STD	Sls LTD	Rec'd LTD	On Hand	On Order	ST% TW	ST% STD	ST% LTD	Pen% TW	Pen% STD	Pen% LTD	Pen% OH
Virtual Style: B0773040 SP08																
D:149 C:85 SP08 RACE	$525.00	$220.00	12/26/07	0	14	18	45	27	0	0.0%	34.2%	40.0%	0.0%	21.5%	22.0%	8.9%
V:51859 S:RACE 001:BLACK			12/26/07	$0.0	$7.2	$9.3	$23.6	$14.2	$0.0	0.0%	33.7%	39.7%	0.0%	24.5%	24.8%	9.8%
D:149 C:85 SP08 RACE	$525.00	$220.00	12/26/07	0	1	2	18	16	0	0.0%	5.9%	11.1%	0.0%	1.5%	2.4%	5.3%
V:51859 S:RACE 211:TAN			12/26/07	$0.0	$0.5	$1.1	$9.5	$8.4	$0.0	0.0%	5.9%	11.1%	0.0%	1.7%	2.9%	5.8%
D:149 C:85 SP08 ROLLER	$525.00	$220.00	12/26/07	0	4	7	54	47	0	0.0%	7.8%	13.0%	0.0%	6.2%	8.5%	15.6%
V:51859 S:ROLLER 001:BLACK			12/26/07	$0.0	$2.0	$3.5	$28.4	$24.7	$0.0	0.0%	7.4%	12.6%	0.0%	6.8%	9.3%	17.1%
D:149 C:85 SP08 ROLLER	$525.00	$220.00	12/26/07	0	4	4	42	38	0	0.0%	9.5%	9.5%	0.0%	6.2%	4.9%	12.6%
V:51859 S:ROLLER 211:TAN			12/26/07	$0.0	$2.1	$2.1	$22.1	$20.0	$0.0	0.0%	9.5%	9.5%	0.0%	7.1%	5.6%	13.8%
D:149 C:85 SP08 SOSO	$395.00	$185.65	12/26/07	1	13	16	40	24	0	4.0%	35.1%	40.0%	50.0%	20.0%	19.5%	7.9%
V:51859 S:SOSO 100:WHITE			12/26/07	$0.4	$5.0	$6.2	$15.8	$9.5	$0.0	4.0%	34.7%	39.6%	46.0%	17.0%	16.5%	6.6%
D:149 C:85 SP08 SOSO	$395.00	$185.65	12/26/07	0	2	3	24	21	0	0.0%	8.7%	12.5%	0.0%	3.1%	3.7%	7.0%
V:51859 S:SOSO 400:AZZELA			12/26/07	$0.0	$0.8	$1.2	$9.5	$8.3	$0.0	0.0%	8.7%	12.5%	0.0%	2.7%	3.2%	5.7%
D:149 C:85 SP08 SOSO	$395.00	$185.65	12/26/07	0	5	6	24	18	0	0.0%	21.7%	25.0%	0.0%	7.7%	7.3%	6.0%
V:51859 S:SOSO 600:FRESA			12/26/07	$0.0	$1.6	$2.0	$9.5	$7.1	$0.0	0.0%	18.2%	21.7%	0.0%	5.4%	5.3%	4.9%
D:149 C:85 SP08 SPIRE	$475.00	$198.00	12/26/07	1	10	12	47	35	0	2.8%	22.2%	25.5%	50.0%	15.4%	14.6%	11.6%
V:51859 S:SPIRE 001:BLACK			12/27/07	$0.5	$4.6	$5.6	$22.3	$16.6	$0.0	2.8%	21.8%	25.1%	57.5%	15.6%	14.9%	11.5%
D:149 C:85 SP08 SPIRE	$475.00	$198.00	12/26/07	0	2	3	43	40	0	0.0%	4.8%	7.0%	0.0%	3.1%	3.7%	13.2%
V:51859 S:SPIRE 200:BROWN			12/26/07	$0.0	$0.8	$1.3	$20.4	$19.0	$0.0	0.0%	4.2%	6.4%	0.0%	2.7%	3.5%	13.1%
D:149 C:85 SP08 TWO-TONE L/U	$475.00	$198.00	12/26/07	0	10	11	47	36	0	0.0%	21.7%	23.4%	0.0%	15.4%	13.4%	11.9%
V:51859 S:TWISTER 202:CHOC/CREAM			12/26/07	$0.0	$4.8	$5.2	$22.3	$17.1	$0.0	0.0%	21.7%	23.4%	0.0%	16.3%	13.9%	11.8%
TOTAL FOR VIRTUAL STYLE: B0773040 - SP08				2	65	82	384	302	0	0.7%	17.7%	21.4%				
				$0.9	$29.4	$37.5	$183.3	$144.8	$0.0	0.6%	16.9%	20.6%				
Virtual Style: B0773041 FL07 & PRIOR CONT																
D:149 C:85 SP07 SOSO	$395.00	$185.65	06/13/07	0	12	14	38	24	0	0.0%	33.3%	36.8%	0.0%	22.2%	9.6%	12.4%
V:51859 S:SOSO 001:BLACK			12/26/07	$0.0	$4.6	$5.4	$15.0	$9.5	$0.0	0.0%	32.9%	36.4%	0.0%	21.7%	9.6%	11.1%
D:149 C:85 SP07 SOSO	$395.00	$185.65	06/14/07	0	10	41	70	29	0	0.0%	25.6%	58.6%	0.0%	18.5%	28.1%	14.9%
V:51859 S:SOSO 024:DARK GREY			12/26/07	$0.0	$3.9	$14.0	$27.7	$11.5	$0.0	0.0%	25.2%	54.9%	0.0%	18.4%	24.9%	13.4%
D:149 C:85 SP07 SOSO	$395.00	$185.65	03/15/07	0	2	6	11	5	0	0.0%	28.6%	54.6%	0.0%	3.7%	4.1%	2.6%
V:51859 S:SOSO 105:GREEN			03/15/07	$0.0	$0.7	$1.9	$4.3	$2.0	$0.0	0.0%	26.5%	49.6%	0.0%	3.3%	3.4%	2.3%
D:149 C:85 SP07 SOSO	$395.00	$185.65	06/15/07	0	12	27	48	21	0	0.0%	36.4%	56.3%	0.0%	22.2%	18.5%	10.8%
V:51859 S:SOSO 200:BROWN			06/15/07	$0.0	$4.7	$9.5	$19.0	$8.3	$0.0	0.0%	36.0%	53.5%	0.0%	22.2%	16.9%	9.7%
D:149 C:85 SP07 SOSO	$395.00	$185.65	12/26/07	0	5	8	45	37	0	0.0%	11.9%	17.8%	0.0%	9.3%	5.5%	19.1%

3.11 Sell-through report from a U.S. department store

A sell-through report is a sales report generated by retailers that lists each style, quantity, location, and rate of sale. This allows the buyers to assess how popular and profitable each product is and will help with reorders and future budget plans.

3.12

" THE MAJORITY OF BRANDS OFFER CORE STYLES, WHICH IS REALLY IMPORTANT AS VOLUME DRIVERS. KEY PIECES INCLUDE THE CLASSIC BB PUMP AND THE HANGISI BEJEWELLED SATIN PUMP FROM MANOLO BLAHNIK, CLASSIC BALLERINA FLATS FROM LANVIN, AND THE ROCKSTUD KITTEN-HEEL PUMPS FROM VALENTINO. FASHION STYLES ARE LAYERED IN ON A SEASONAL BASIS TO ADD NEWNESS AND FASHION ELEMENTS, AND THIS IS THE MAIN FOCUS FOR PRESS AND MARKETING."

JASMIN SANYA, FOOTWEAR BUYER, HARVEY NICHOLS

3.12 Hangisi bejewelled satin pump from Manolo Blahnik

Reworking a classic or best-selling style for several seasons is a key component of range planning. For many well-known brands, the customer will return season after season, year after year to purchase a similar shoe that he or she has had before because it is reliable and fits well. It is essential to offer these key styles, and they make up core lines within a collection.

WHOLESALE DISTRIBUTION

Channels that goods flow through to reach the retailer and ultimately the consumer are both well established and multi-layered. Wholesale distribution covers the reseller market and includes trade fairs for brands and private-label manufacturers to exhibit, off-site exhibitions, and showrooms, as well as agents and sales representatives.

Trade Fairs

Trade fairs serve two main purposes in the footwear trade: as a meeting point for businesses to interact—promoting, buying, and selling products through wholesale distribution—and as a place to identify key trends for the season. Recent years have shown trade fairs evolving in their role as facilitators for the industry, not just as a place to network but also to attend seminars and awards ceremonies. A list of key international trade fairs is in the resources section.

Off-Site Exhibitions

Brands also rent space in hotel suites, corporate meeting rooms, or small temporary showrooms coinciding with fashion weeks in New York, London, Milan, and Paris. Hotels such as the Park Hyatt, Paris, and Soho House, New York, offer exhibition space and hotel suites adapted to show collections. Not always a less expensive option than a trade fair, it allows a more discreet environment for buying appointments or space to exhibit if one is not available at the main exhibition.

3.13 London Fashion Week shoe salon
Key fashion cities such as London include footwear exhibitions on their Fashion Week schedule for new and emerging British footwear brands, allowing them to meet international press and retail buyers.

3.14

Showrooms

Many brands do not show at trade fairs as their export strategy and international distribution may be well established with a number of international or regional showrooms. Jimmy Choo, Prada, Gucci, etc. have showrooms in their regional or head offices. This avoids exposing the new season's collection and ensures the brand's exclusivity and control of retail distribution. Careful consideration is given to new clients and stockists, checking boutique store locations, credibility, and adjacent brands stocked before inviting them to view the collection. There is often a minimum spend required, and the account is expected to keep to this each season regardless of whether it has been successful or not.

3.14 FFANY NYC
Footwear brands exhibiting in international locations such as New York may choose to show in a temporary space such as a hotel suite.

3.15 Lotus exhibiting at FFANY NYC
British brand Lotus exhibiting at FFANY NYC in the Hilton Hotel.

3.15

CASE STUDY:
Bata India Ltd.—Vertical Integration and Controlling Costs

The latter part of the twentieth century saw apparel manufacturers move away from the traditional model of a company that owned all aspects of its operations in a bid to cut costs. However, a new form of vertically integrated company is evolving in footwear. The vertical integration (VI) of a company's operations can offer survival for longstanding manufacturers because the internet offers a direct route to the end consumer. Old and new companies are increasingly able to control their manufacturing, brand identity, and route to market.

Bata

Bata India Ltd. was first established by a Czech, Tomaz Bata, in 1931 and went public on the Indian Stock exchange in 1973. As India's largest footwear manufacturer and retailer, it offers affordable leather, rubber, canvas, and PVC shoes to the Indian market and overseas. There are five factory sites; the first was established in Batanagar. As the market grew, the company expanded its production throughout India, opening factories in different states to meet demand. Bata India also owns two tanneries, one in Batanagar and India's second largest, Mokamehghat, in Bihar. The company now has over 500 wholesale clients and manages over 1,500 retail outlets across India, as well as an online transactional websites globally. Bata India has a selection of sub-brands and annually sells around 60 million pairs of shoes. The company also distributes and markets a number of international brands, such as Scholl and Hush Puppies.

Bata India is both a backward (owning tanneries) and forward (owning retail distribution) vertically integrated company. It manages the entire supply chain by controlling the supply of raw materials, footwear design, and production, and through its distribution channels offers wholesale, export, and own retail through stores and online.

The advantage is overall coordination; many of the holdups in the supply chain, such as a delay in delivery of raw materials or shipping, can be advised and dealt with more quickly, therefore reducing lead times. It also allows Bata India to reduce margins at certain points in the supply chain, as internal costs may be lower. However, being the largest retailer in India, the company may well hold a monopoly over the market and destroy competition quickly, which allows little choice for the consumer.

> " A VERTICALLY INTEGRATED COMPANY IS ONE WHICH OPERATES AT MORE THAN ONE LEVEL OF THE PIPELINE."
> JONES (2006)

Industry Perspective:
John Saunders, CEO, British Footwear Association

John Saunders has spent his career working in the footwear industry, predominantly in global sales management positions for a number of key British and US brands. He is currently the chief executive of the British Footwear Association, which has been committed to promoting British talent and skills within the footwear industry for over a century.

What is your job role and can you describe a typical day?

I may be meeting start-up design companies one day to discuss how they can get going in the industry, advising where to source component or find partner factories, or speaking to one of our established brands about issues affecting their global business. I work with the UK government on export funding proposals and challenges via UKTI (UK Trade and Investment) and with CEC, the European Confederation of the Footwear Industry, on matters such as international duty rates and factory audits. My role brings me into contact with the organizers of some of the biggest trade shows in the world to plan UK pavilions and marketing initiatives;

I represent the UK footwear industry on the Industrial Liaison committees of leading UK universities advising on the commercial content of their course and work with new brands to identify and providing one-to-one mentoring via industry specialist coaches via the BFA's own Accelerated Development Programme.

Describe the importance of trade associations for both established and start-up businesses.

Starting a new company or brand is never easy, and access to the right information and resources is difficult even when you know where to look. Trade associations like the BFA have become expert in providing support, resources, training, and guidance for their members with particular focus on start-ups.

As the UK retail market becomes more concentrated and higher value retail opportunities, especially for new brands, become more limited, UK brands are being driven to seek new business opportunities in overseas markets. Young businesses turn to export as a route to develop their business, and the BFA has recognized the need to develop a program of training and support for its members to provide expert guidance on a wide range of topics and issues that they will undoubtedly encounter on the venture overseas.

Our members utilize our in-house team, external specialist consultants, and workshops for guidance on export pricing, margins, and duties; marketing; range development; legal matters including IP and brand registration; agent and distribution contracts; exhibiting and merchandising; and export finance as well as funding and support.

bfa british footwear association

SUMMARY

Asia, notably China, is the dominant force in all aspects of the footwear trade. Issues around labor costs and currency may impact on China's ability to maintain its low-cost production, which gives room for larger producers of footwear such as India and Brazil to increase their exports. There is also evidence that other Asian countries are in a good position to take some low-cost production from China. Currently, no other country has the population, coordinated government policies, and workforce to rival China as exporters.

Despite various governments introducing trade tariffs and free trade agreements, sourcing fashion footwear follows cheap and flexible production to meet consumer demand. Low-cost production declines where labor costs increase.

The footwear buyer has a complex role in today's globalized supply chain with responsibilities to source ethically, legally, and quickly while coordinating many of the key activities that are essential for success. Brand owners and buyers operate in a competitive field with constant pressures to reduce or increase margins to make a profit, and as such companies look to adapt and modify their business models.

DISCUSSION QUESTIONS

1. Can importing cheap shoes directly lead to the reduction or demise of footwear manufacturing in a country? Discuss the pros and cons.

2. Can duties and raising taxes hinder or stop retailers from importing product from overseas?

3. How does this affect manufacturers and the consumer in the countries you are familiar with?

4. Which trade association supports the footwear industry in your country?

5. What are the most important considerations when working out wholesale to retail prices?

EXERCISES

1. Highlight, analyze, and explain the specific issues that affect fashion footwear retailers and brands when:
   a. sourcing and maintaining supplier relationships from various regions, and
   b. managing consumer perceptions of the brand within the UK, United States, etc.

2. What are the advantages and disadvantages of vertical integration for a start-up company?

3. In light of globalization, how might a well-established manufacturer consider adapting its working practices to a more VI approach?

KEY TERMS

Anti-dumping duties—also referred to as quotas, as they are a tax imposed to restrict free flow of low-cost items that may threaten an existing manufacturing base in the import country

Break-even—point at which sales revenue covers all the fixed and variable costs in a company, after which a profit can be made

Budget—a set financial plan/amount of money that can be spent on new stock and reorders

Comparative shopping—identifying a competitor's product range, prices, and promotional behavior by visiting its store and making online observations

Cost price—the amount of money paid for a product from the manufacturer (with the intention of reselling for a profit)

Credit terms—an agreement between buyer and seller that states the buyer will pay for the goods a set number of days after they have been received

Discounts—money off for early payment of invoices or bulk purchases

Economies of scale—costs saved by increasing production output of shoes, making the average price per pair cheaper

Electronic point of sale (EPOS)—the collation of number of units and price of goods sold at each retail location

(Free) trade agreements—arrangements made between countries that import and export with each other at favorable, low, or no extra taxes

Incoterms—set of instructions detailing who is responsible for each stage of shipment

Logistics—the management and movement of goods along the supply chain

Margins—a percentage of the selling price that covers running costs and profit (excludes the cost price)

Markdown—a reduction in the retail price to incentivize sales

Markup—doubling or tripling of the cost price of goods

Merchandiser—manages the movement of products and prices to maximize sales and profit in the retail environment

Open to buy—budget available to buy new or reorder products within a current season

Range planning—the selection of complementary products for retail which considers price and customer

Sales projections—forecast of how many products need to be sold at retail to make a profit

Sell through—amount of product that is sold in a given time frame

Selling price—amount of money the goods are sold for in the retail store

Stock turn—number of times a product is sold in a given time frame

Trade deficit—when a country's imports exceed the value of its exports

Value added tax (VAT) or sales tax—government tax levied on the consumer at the point of sale

Vertical integration—when a company owns several different operations along the supply chain

Wholesale distribution—sale of goods in bulk to be resold through retail outlets to the end consumer

04

THE RETAIL AND E-TAIL LANDSCAPE

Learning Objectives

• Explain the difference between private-label and branded product, and outline the different retail channels where both types of product are sold across market levels.

• Discuss how different retail brands employ different modes of market entry and either a push or pull marketing strategy in conjunction with a multichannel or omni-channel approach.

• Identify and discuss new retail models emerging from an omni-channel approach.

• Evaluate the need for both traditional and new, "inventory-less" retail models.

4.1 Iconic luxury brand Christian Louboutin Boutique at Harrods, London
Harrods re-launched its shoe department with footwear partner Kurt Geiger in 2014 with "Shoe Heaven," comprising seventeen concessions and over fifty designers across 42,000 square feet.

INTRODUCTION

In this chapter, key retailers are explored in some of the most significant marketplaces for fashion footwear, identified by the largest numbers of pairs purchased per capita. These markets have been identified as the United States and the UK, as both markets have the highest number of pairs purchased per head (via The World Footwear Yearbook, http://www.worldfootwear.com). Case studies demonstrating the changing face of retail and e-tail are examined, within different product sectors and at different market levels, highlighting the challenges that shoe retailers face today.

The challenges that footwear retailers have always faced, e.g., stockroom space versus retail space and creating a comfortable environment to try shoes on, continue to test retailers to arrive at more creative solutions regarding use of expensive real estate. This shift and change has been exacerbated not only by high real estate costs but also by the success of online as a channel. Retailers are now adopting a multichannel, or rather omni-channel, approach to maximize sales and attract new consumers.

As consumers in the dominant and influential markets become more discerning, diverse, and knowledgeable, they are actively looking for a point of difference—not just with their product preferences but by their increased expectations as to what makes a seamless and positive consumer experience. This chapter explains how and where consumers purchase their favorite fashion footwear and also highlights some new and emerging retail models that see the lines between retail, wholesale, and media blurring.

RETAIL AND PRODUCT CLASSIFICATIONS FOR FASHION FOOTWEAR

Fashion footwear is sold through a multitude of retail channels, across market levels. One of the most significant factors is whether the product is branded, i.e., designed, produced, and retailed in either the brand's "own doors" and via wholesale, or "private label" (also referred to as "own label") and is sold exclusively through the retailer's own retail outlets.

NONSPECIALIST / SPECIALIST

PRIVATE LABEL

Supermarkets, e.g., Walmart and Tesco (F&F fascia)
Fashion multiples, e.g., Forever 21, Eileen Fisher, New Look, Zara (Inditex), and TopShop (Arcadia)

Multiple store format, e.g., Skechers and Clarks

BRANDED

Multiproduct brands, e.g., Prada, Gucci, and French Connection

Footwear, e.g., Christian Louboutin, Russell and Bromley, and Brown Shoe Company

Department Stores, e.g., Saks Fifth Ave, Nordstrom, Dillard's, Selfridges, and Harrods

Independent boutiques, e.g., Cricket (Liverpool) and Opening Ceremony (New York, Los Angeles, and Tokyo)

Independent footwear stores, e.g., Gimme Shoes (San Francisco) and Pam Jenkins (Edinburgh)

4.2 Branded and private-label product at retail, Fiona Armstrong-Gibbs
Branded and private-label product is sold at different market levels across different retail channels and can be categorized as retailers that sell footwear only (specialists) and multiproduct stores (nonspecialists); examples here are from the United States and UK fashion footwear markets.

4.2

4.3

RETAIL CHANNELS AND KEY PLAYERS

The largest market in terms of pairs purchased per capita is the United States; therefore, it has the most developed fashion footwear market because consumers are purchasing product for aesthetic rather than purely practical needs. Sports, casual, and lifestyle brands occupy the highest market shares, with Nordstrom Inc. and Dillard's Inc. representing the luxury and premium market levels within the department store channel.

The U.S. footwear market reached $52.59 billion in 2014. In 2019, footwear retail in the United States is forecast to reach $60.53 billion, representing a value Compound Annual Growth Rate) (CAGR) of 3.1% since 2015 (Mintel, 2015).

It is estimated that women's products accounted for 48 percent of sales, while men's and children's accounted for 35 percent and 17 percent, respectively. The women's category has shown the greatest revenue growth in the last five years, at $3.2 billion. However, the fastest-

Market Player	2013	2014
Nike, Inc.	12.8	14.1
Wolverine World Wide, Inc.	6.8	6.7
Wal-Mart Stores, Inc.	5.9	5.8
Nordstrom Inc.	5.3	5.2
Brown Shoe Company	4.1	4.1
New Balance Athletic Shoe, Inc.	3.7	4.2
Kohl's Corporation	2.8	2.8
Skechers	2.4	2.9
Authentic Brands Group, LLC (Jones Apparel Group, Inc)	2.1	1.9
Dillard's Inc.	2.1	2.0
Ross Stores Inc.	2.1	2.3
Others	49.9	48.0

4.4

4.3 US footwear: company retail market share by value (percent)
This data includes men's, women's, and children's footwear through all retail outlets, including direct to consumer; it excludes industrial/work footwear but includes sports and casual shoes and boots. Source: © 2016 Mintel Group Ltd. All rights reserved.

4.4 Skechers "Go Walk" lifestyle product, Macy's, New York
Skechers retail across multiple retail channels, such as department stores, sporting goods stores, and independent stores.

growing category is children's footwear, having grown 17 percent in the same period (2010–2014). In recent years there has been a shift where athletic-inspired and comfort styles have gained traction as sports and lifestyle merge.

Sports and lifestyle brands dominate the market where "Casual Friday" has become commonplace every day.

General merchandise stores include both mass-market discounters and department stores, e.g., Macy's and Nordstrom, and national chains such as Sears and JC Penney.

Sports and lifestyle brands and styles can be found across all retail channels as this is increasingly the most popular consumer choice.

The UK footwear market has grown from £9.6 billion in 2014 to $10.3 billion in 2015 and is forecast to grow to £13.4 billion in 2020. In 2014, women's shoes represented 55 percent of the market, with men's and children's shoes at 35 percent and 10 percent, respectively.

4.5

		2012 %	2013 %	2014 %
1	C&J Clark	8.5	8.5	8.2
2	Sports Direct	6.1	6.6	6.8
3	JD Sports Fashion	5.0	5.0	5.5
4	Marks & Spencer	5.7	5.4	5.1
5	New Look	4.4	4.1	4.2
6	Primark	3.6	3.7	3.8
7	Office	3.2	3.7	3.8
8	Next	3.8	3.6	3.4
9	Kurt Geiger	2.9	2.9	3.0
10	Schuh	2.9	2.8	2.8
11	Shop Direct Group	2.7	2.7	2.7
12	Shoe Zone	3.1	2.6	2.2
13	Macintosh Retail Group (Netherlands) Of which:	2.4	2.3	2.1
	A Jones & Sons	1.0	1.0	0.9
	Brantano	1.4	1.3	1.2
14	Dune Group	2.1	2.1	2.1
15	TK Maxx	1.8	1.8	1.8
16	Russell & Bromley	1.5	1.5	1.6
17	Debenhams	1.6	1.5	1.4
18	Asda	1.5	1.4	1.4
19	Hotter/Beaconsfield Footwear	1.0	1.2	1.3
20	Matalan	1.2	1.2	1.2

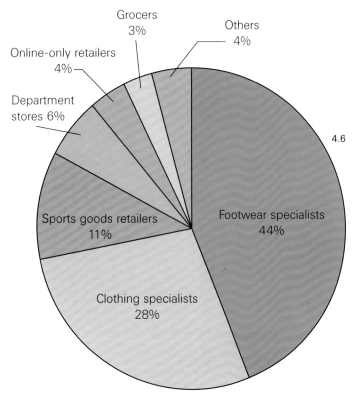

4.6

4.5 UK footwear market: company retail market share by value (percentage)
This data includes all footwear, including trainers, but excludes slippers and specialist performance shoes. © 2016 Mintel Group Ltd. All rights reserved.

4.6 Distribution of UK footwear spend by retailer type, 2015 (estimated)
Footwear specialists have been challenged by multiproduct retailers who recognized the opportunity in the growing market. © 2016 Mintel Group Ltd. All rights reserved.

4.7

FOOTWEAR SPECIALISTS

While recent years have seen growth in nonspecialist retailers in general, this has prompted specialist retailers to promote their expertise in fitting, customer service, and after sales to create a point of difference rather than competing on price. Founded in 1825 by Quaker brothers Cyrus and James, C. & J. Clark International Ltd. in Street, Somerset, Clarks operate out of thirty-five countries via its own branded stores and also via wholesale, selling more than 50 million pairs per year. With over 1,000 branded stores globally, the company aims to open another 100 stores during the next five years. It is focusing on franchising to benefit from local expertise in new territories in order to reach this target.

Clarks has responded to challenging economic climates throughout its 190-year history by maintaining its high standards of customer service and focus on staff training to fit shoes correctly in-store, providing continuity in sizing that creates consumer confidence to order online, encouraging repeat purchases online.

4.8

4.7 Kurt Geiger, Heathrow, London, UK
Fashion footwear specialist Kurt Geiger sells branded and private-label product from department stores and its own stores, including airport retail, such as Heathrow, London; "traffic stopping" displays invite consumers in.

4.8 Clarks outlet store in Bicester Village, UK
Clarks offers many styles in wide and narrow fittings in both its full-price and outlet stores; this focus on comfort through correct fitting has been an enduring feature of this retailer.

Nonspecialist retailers have now surpassed specialists, who in 2014 had 44 percent market share.

The UK market mirrors a similar pattern to the United States in that comfort, sports, and lifestyle retailers dominate. Nonspecialist retailers recognized the opportunity in this product category at the start of the recession and have made fashion footwear a strategic priority; this has been especially successful at the value and fast fashion end of the market. The only retailer that represents the premium market within the top ten footwear retailers is Kurt Geiger.

Multiples

Multiples (also known as chain stores) operate out of multiple locations and have common ownership and management. They can sell "private label" (also known as "own label") product—for example Clarks and Sketchers—or offer a combination of their own (private-label ranges) and branded product such as Russell and Bromley, which sells its own label and brands such as Stuart Weitzman, Sebago, and Barker.

Independents

These specialist stores are small stores with a single location (or up to three stores) and can be owned by an individual, family, or a two-person partnership. Independents are often off the beaten track (not on the high street due to retail costs) and cater to niche consumer groups who will seek these stores out for specific styles and brands. For example, Gimme Shoes, established in San Francisco in 1984, sells men's and women's designer footwear and accessories and has three stores in the city. It has garnered a loyal following by offering fashion-forward premium designer brands in a relaxed environment and focusing on service.

NONSPECIALIST RETAILERS

These operate at multiple market levels; from supermarket giants such as Walmart and Tesco's F&F label to fashion multiples whose primary revenue comes from clothing, such as Forever 21, Eileen Fisher, New Look, Zara (Inditex) and TopShop (Arcadia). Due to their buying power and therefore ability to negotiate highly competitive cost prices, these retailers pose a threat to the business of footwear specialists. They can produce in vast quantities under their own private label or sub-brands sold within their own stores. This product can only be found within their "own" fascias. They will often collaborate with designers/designer brands that operate at a higher market level to enhance their credibility, such as Hennes and Mauritz working with Isabel Marant and clog brand Swedish Hasbeens. (Collaborations are discussed in detail in Chapter 8.)

4.9 Global luxury footwear—the world's largest footwear destinations (mixed sources)
The race to create the world's largest footwear destinations is highly competitive among luxury department and specialist stores.   4.9

	Location	Square footage	Opening date	Additional information
Level Shoe District	Dubai	96,000	2012	300 designers, 40 boutiques, with additional services such as repairs, stylists, pedicures, therapies, and a concierge
Macy's	Herald Square, New York	63,000	2012	430 dedicated footwear-only employees and 280,000 pairs of shoes stocked
Harrods	Knightsbridge, London	42,000	2014	Coined "Shoe Heaven," the area has 150 sales associates, 50 designers, 17 concessions, and 100,000 pairs of shoes stocked
Selfridges	Oxford Street, London	35,000	2010	4,000 shoes on display, with 55,000 pairs stocked and 11 designer boutiques/concessions
Saks	5th Avenue, New York	17,500	2007	Saks worked with the U.S. Postal service to designate it its own zip code, 10022-SHOE

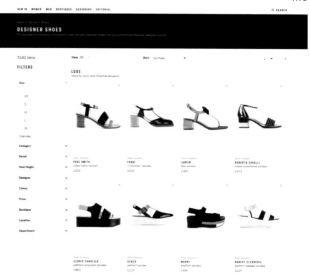

4.10 Farfetch women's premium product
A positive brand image can be maintained via well-curated and -presented transactional pages at premium level, such as Farfetch.

Multiproduct "Designer" Brands

These brands offer footwear as a part of their overall product mix, and while footwear is not necessarily a specialism, due to the luxury or premium nature of their offer, the same high standard is usually reflected in the quality of their footwear.

This kind of product is retailed via "own" doors (flagship stores and their single fascia stores), as well as being sold via wholesale to department stores and independents via the brand's own sales team, distributors, or agents (see modes of market entry), e.g., Prada, Gucci and French Connection.

Department Stores

These large retail outlets are situated in key locations in city centers and are important "anchor" stores for malls selling multiple product categories from an assortment of brands. They can be privately owned independent businesses (such as Harrods) or part of a larger group (such as Nordstrom or Macy's).

Shoe halls are highly competitive areas in the international footwear arena with great pride and prestige in being associated with having some of the world's largest retail areas devoted purely to footwear.

E-TAILING

E-tailing is the selling of retail goods on the internet. It is a subset of e-commerce, which encompasses the buying and selling of goods and services via the internet.

At the start of the e-commerce era there were many skeptics regarding how successfully footwear could be sold online; however, this is a growing sector today. The negatives (such as not being able to try footwear on) are outweighed by the convenience for the consumer, although higher returns rates online pose different challenges for retailers and brands.

The average price of fashion footwear purchased online is higher than that in store, which suggests that consumers are inclined to invest in branded footwear as a destination online purchase more so than in-store. This would also suggest that that the lure of a bargain is more powerful in-store rather than online. Mid-market and premium e-tailers are now firmly established as trusted "go-to" sources through companies such as Zappos, Sarenza, and Javari (Amazon) as well as boutique buying groups such as Farfetch.

Pure Play E-tailers

In e-tail terms, pure play businesses sell solely via the internet and have no physical retail stores (brick and mortar stores). Footwear specialist pure play e-tailers include Shoescribe.com and Rubbersole.co.uk, and nonspecialists include Net-a-Porter and Amazon.

MULTI- AND OMNI-CHANNEL RETAIL

Many successful retailers and brands use a combination of approaches to fully maximize sales opportunities and reach the optimum number of consumers. What was known as a multichannel approach, i.e., selling product via a variety of retail channels, such as brick-and-mortar stores as well as online (retail and e-tail), has evolved into "omni-channeling," whereby a seamless consumer experience is created across all available shopping channels, e.g., mobile internet devices, computers, brick-and-mortar, television, direct mail, and catalogues.

Tory Burch employs a multichannel retail approach in that her product is available from her own stores (flagship stores), department stores, and online via her own transactional website. This way, the distribution is controlled and managed as the company expands into other product areas.

A "push" marketing strategy is used to drive traffic to the stores or to purchase via the company website.

4.11

4.12

4.12 Dr. Martens "WHAT DO YOU STAND FOR? #STANDFORSOMETHING" window display and retail POS
Dr. Martens challenges the consumer to respond to this question and enter into a dialogue with the brand via the #STANDFORSOMETHING campaign.

Dr. Martens favors an omni-channel approach and sells via it's own stores (flagship stores), footwear multiples, independent footwear specialists, department stores, and its own transactional website. It uses a "pull" marketing approach whereby its diverse consumer demographic is invited to enter into a dialogue with the company via interactive campaigns and in-store point of sale (POS), such as its "What do you stand for" campaign.

M-Commerce

Mobile commerce is one of the fastest growing retail channels; it is the buying and selling of goods and services through wireless handheld devices such as cellular phones and personal digital assistants (PDAs). Known as next-generation e-commerce, m-commerce enables users to access the internet without needing to plug in.

4.11 Tory Burch flagship store, Meatpacking District, New York City
A premium brand image is well maintained and enhanced at Tory Burch flag ship stores.

MODES OF MARKET ENTRY

If a retailer and/or brand are entering a new market, there are many different business models that they may choose to adopt in order to build a presence in a new geographical territory. Large, established retailers/brands and upcoming designers alike investigate these options and make a decision based upon their financial position and the micro and macro environment (as described in Chapter 5).

It is not always practical or profitable for the same business model to be used in the international market as the domestic (home) market, so retailers and brands will investigate the most appropriate route for their international expansion via some of these methods:

• Joint venture (J.V.)—a contractual business agreement between two or more parties. It is similar to a business partnership, although more usually is based upon a single business transaction or project rather than a longer-term partnership. Advantages are that strengths and expertise are shared, minimizing risks to increase the competitive advantage in the new marketplace, for example Spanish brand Mango for JC Penney (retailing as MNG).

• Concession—where a brand or designer operates an area within a larger store or retail area, most usually a department store. The company identity and ethos is upheld by the area being run as if it were a separate entity. The "host" store receives a premium and rent and will benefit from the extra customers that the area attracts. Contractual aspects such as who pays for staff or contributions to staff costs will vary; however, it is necessary that there are strong reciprocal benefits for both parties, as it does for Michael Kors concessions at Harrods and also at airport retail destinations such as Singapore Airport and JFK, New York.

• Franchise—when a brand/retailer gives permission to another business entity to sell its product on its behalf in another geographical territory. This company must sell the goods in an identical way to the retailer's domestic stores, and the franchisee pays back a percentage of sales to the retailer, often in return for some element of exclusivity within the new marketplace. For example, Alshaya operates a Harvey Nichols franchise in Kuwait.

• Flagship/own retail—where no other party is involved, other than a real estate company or property agent if the retail premises are not fully owned. An example is the Nicholas Kirkwood store in the Meatpacking District, New York City.

4.13

4.13 Michael Kors concession, Macy's, New York City
Michael Kors concessions in department stores and airports make excellent use of a limited amount of space while still retaining the brand's premium image.

- Wholesale-Distributor—a company that buys noncompeting products in bulk from multiple and complementary brands, and then resells them either to local retailers or to consumers directly, via brick-and-mortar stores and online. Distributors in a new market will often have a knowledge and expertise in the product sector as well as established warehousing, freight, marketing, and manpower capabilities. For example, Viva Trading is the UK distributor for US brand Minnetonka Moccasins.

- Wholesale-Agent—an agent or agency will act on a brand's/retailer's behalf and sell its product on a percent commission basis. Agents work with similar and noncompeting brands and carry several of these ranges at a time and work directly with retail buyers. Established agents usually have their own premises, whereas upcoming agents/agencies will visit the retailer's premises. Agents will secure exclusivity within a geographical territory via a contractual agreement with the designer/retailer. For example, RainbowWave (which has showrooms in Paris, New York, and London) sells Ancient Greek Sandals and Rachel Comey.

- Wholesale-Subsidiary—when an enterprise is controlled by another (called the parent company) via the ownership of more than 50 percent of its voting stock; for example, Deckers UK was set up as a subsidiary of US-based Deckers Corporation after AMG Group, the former distributor of UGG Australia, handed back responsibility and management of Ugg Australia to Deckers in 2010. The London-based regional office is directly responsible for managerial functions for the UK market, including sales, marketing, finance, and operations.

- Licensing—a business agreement usually bound by contracts for a fixed term whereby the brand (intellectual property owner or licensor) gives another party permission to design, manufacture, and sell product bearing the brand's name (as a registered trademark). The licensee usually specializes in this product sector, having an expertise that the brand owner does not have, so the licensee pays the intellectual property owner a royalty, e.g., a percentage of sales. For example, Brazilian manufacturing company Grendene SA is the licensee for Hello Kitty and Disney children's footwear.

4.14 Nicholas Kirkwood flagship store, New York City
Nicholas Kirkwood flagship store, Meatpacking District, New York City.

THE CHANGING FACE OF RETAIL

Retail is evolving to provide increasingly knowledgeable, discerning, and demanding consumers with more choice than ever before. As such, the traditional notions of established retail models are being challenged to provide the consumer with the choice and variation that they crave.

One such solution is STORY in New York (Chelsea). Every month to six weeks, shoppers are invited to experience different themes in-store, offering carefully curated limited-edition product across fashion and lifestyle product sectors, catering for men, women, and kids. Owner Rachel Shechtman previously worked as brand consultant (for brands such as Toms footwear) and has adopted an editorial approach as inspiration for this business model. Not to be confused with a pop-up shop, this is a permanent store with a unique approach that is based on the following two fundamental hypotheses:

1. The future of retail will still be about consumption, but more about having a community and the shared experience they have.

2. Retail is an untapped frontier for advertising, and there's an intimate opportunity for conversation between the consumer and brand. (Source: http://www.forbes.com/sites/vannale/2012/02/09/love-and-retail-a-nyc-boutique-redefines-the-shopping-experience/#3245ece32161)

4.15 STORY, New York City; FitFlop and Stephen Jones collaboration, limited edition
Limited-edition collaboration footwear is cross-merchandised as part of the "Cool Story" theme at STORY, New York City.

From a footwear perspective, footwear features as part of the product mix, and the product changes completely every four to six weeks so that the customer always has a reason to return on a regular basis. The store was profitable within two years, due to its unique point of difference.

> " STORY HAS THE POINT OF VIEW OF A MAGAZINE; EVERYTHING CHANGES EVERY FOUR TO EIGHT WEEKS LIKE A GALLERY, AND IT SELLS THINGS LIKE A STORE. A MAGAZINE TELLS STORIES BETWEEN PICTURES AND WRITTEN WORDS, AND WE DO IT THROUGH MERCHANDISING AND EVENTS. AND OUR VERSION OF PUBLISHING IS SPONSORSHIP. "WHEN YOU LOOK AT THE SQUARE FOOTAGE CERTAIN BRANDS OCCUPY, AND YOU LOOK AT THE AMOUNT OF TIME CONSUMERS SPEND IN THOSE SPACES, WHY ISN'T THAT A MEDIA CHANNEL?"
> RACHEL SHECHTMAN, STORY'S OWNER

STORY's media retail model is based on these seven defining principles:

- CONTENT (product)
- COMMERCE (brand sponsorship to rent space in store provides additional income to retail revenues)
- COMMUNITY (such as educational and social enterprise activities)
- EXPERIENCE (workshops and events)
- EDITORIAL (via the retail design and visual merchandising)
- METRICS (profit achieved after two years due to this innovative model)
- DISCOVERY PLATFORM (a media- and tech-savvy destination)

ETHICS IN ACTION:
The "Minimum" and "Living" Wage

Retail is a major contributor to both the US and UK economies; however, fair living wages that are reflective of real living costs at local level continue to be an issue, and they have been thrust into the limelight in the run-up to the 2016 and 2015 elections, respectively. This is not a new phenomenon with the national minimum wage, which was introduced in the UK by Tony Blair's Labour government in 1998, helping to tackle some of the worst examples of poverty pay by providing a statutory floor. However, post-1998, many UK retailers have used the national minimum wage as the "going rate" rather than being the basis for the minimum starting point.

The retail sector, known for its pressures on margins, is one in which low pay is particularly prevalent. And fashion retail is no exception, as large multiples may exploit the "glamour" of working in fashion. This tends to be less prevalent with independent retailers, where they have a standing and a reputation in the local community; their trade is reliant on local perceptions of them, and unfair treatment of their employees would have a rapid knock-on effect to their business.

In the United States, the federal minimum hourly wage has stayed at $7.25 since 2009 despite Barack Obama urging Congress to introduce a $10.10-an-hour minimum. However, 29 states and the District of Columbia pay a higher minimum wage. Seattle's mayor, Ed Murray, is increasing the city's minimum to $15, the highest in the country's history, phased from 2015–2022.

Walmart's decision in 2015 to increase the wages of 500,000 staff members to at least $9 (£5.80) an hour ($1.75 above the federal minimum,) from April 2015, has been commented on by analysts as driving changes in legislation: "What I've been telling people is that Walmart just raised the federal minimum wage," (Maryam Morse, retail practice leader at Hay Group).

The responsibility resides with consumers as well as retailers. Rhys Moore of the Living Wage Foundation in the UK states, "We hope we'll get to a place where people will start to ask about pay and take it into account when they're shopping." (The living wage in London was £9.15 per hour versus £7.85 out of the capital in 2015.)

CASE STUDY:
Sneakerboy, Melbourne, Australia

Sneakerboy is an award winning "inventory-less" specialist sneaker retailer that opened in Melbourne in 2013. It stocks premium and luxury sneakers from brands such as Rick Owens and Maison Martin Margiela and carries limited-edition "quick strike" product and collaborations such as Raf Simons for Adidas.

Its unique business model is what really sets it apart from other footwear retailers; both its Melbourne and Sydney stores operate without stocking inventory, where consumers can select and try on in store (as well as being to purchase online). The goods are then shipped directly to the customers from the company's warehouse in Hong Kong. Transactions are completed via a Sneakerboy smartphone app or via the store's iPads.

Not only does the business benefit from not using valuable square footage for stockroom space (which can be anything from 30 percent to 60 percent of a footwear retailer's total square footage), but it utilizes the consumer's handheld digital technology to complete the transaction, rather than putting sales through a till point in-store. From a retail perspective, this facilitates higher sales densities and helps manage staffing costs, as well as allows for a really interactive experience whereby the customer can focus purely on the product and is in control of the transaction.

Sneaker enthusiasts at this market level can frequently spend thousands of dollars in one transaction, and having the product delivered two to three days later is often more convenient to this consumer who travels extensively, making carrying multiple shoe boxes impractical and cumbersome.

The Melbourne store was designed by architects March Studio and reflects the store's forward-thinking approach with a futuristic, high-tech interior. This retail concept targets a new generation of luxury consumers with a highly covetable selection of sneakers and an innovative digital retail model that blends the tactility of a physical store with the efficiencies of the internet.

4.16

With 400–500 styles stocked in less than eighty square meters, this is a highly efficient use of space. There are plans to open a store in New York in the future; this looks set to be a retail business model that has great potential for different kinds of footwear going forward and could work well for different product categories.

> " IT'S FUNNY, BECAUSE RIGHT FROM THE BEGINNING WE DIDN'T HAVE A HUGE AMOUNT OF RESISTANCE TO THE ACTUAL MODEL. ONE OF THE THINGS THAT SURPRISED US WAS THAT PEOPLE WERE, FROM THE BEGINNING, VERY ACCEPTING OF THE WAY THAT WE WERE PRESENTING THE OFFER. NOW IT'S NOT EVEN A TALKING POINT; IT'S JUST BUSINESS ON A DAY-TO-DAY LEVEL, THAT'S IN TERMS OF THE RETAIL MODEL."
> CHRIS KYVETOS, OWNER SNEAKERBOY

4.16 Sneakerboy store exterior, Melbourne, Australia
The Sneakerboy circular doorway provides access to the interior, which has been inspired by an underground train station.

Industry Perspective:
Mary Stuart, Owner of mo Brog, Brighton, UK

mo Brog is an independent specialist footwear store in Brighton, UK. Its name is taken from Scottish Gaelic for "my shoe." The store is owned by Mary Stuart, who looks after all aspects of the day-to-day running of the store, including the buying, merchandising, visual merchandising, marketing, PR, and financial planning.

What was your first role in footwear?

Opening mo Brog in March 2013 was my first step into the footwear industry.

What training/qualifications/experience did you have for this role?

I have a strong background in retail, being a Body Shop franchisee for over a decade; I kept my eye on shoes and thought it to be a growing part of the fashion world with a gap in the mid-range market.

I had overlooked until I was quite far into the whole process that my grandfather was a cobbler and a retailer of leather goods. I completed a shoe fitter's course with The British Society of Shoe Fitters where I learnt much about not only shoes but the industry. I also completed a course with Prime, Princes Trust for mature Enterprise, "Preparing to run your own business," which really helped me focus and brush up on marketing.

What is the best thing about running your own store?

The most enjoyable part of my business is putting together a collection every season which brings a fresh start and a new look to the shop. On a day-to-day level I enjoy my customers; the daily interaction is fun, it's very satisfying to see a customer leave the shop with a great pair of shoes that are going to fulfill their desires and needs.

What is the most challenging thing about running your own store?

The most challenging aspect, besides the unpredictable weather, must be suppliers not meeting delivery deadlines, which means launching a new season is tricky, along with the negative financial effect of fewer weeks to sell the stock in. Being a small and new player makes it hard to negotiate a fair discount.

Buying from suppliers who have inconsistent quality control and design issues is problematic; having to return a whole run of shoes or, worse still, customers having to return flawed footwear is bad for business.

In your opinion, which fashion footwear brand is the most iconic and why?

An iconic designer has to be Chie Mihara. Her designs are stunning, flamboyant, adventurous, feminine, and massively eye-catching. They are also incredibly well structured. Having studied orthopaedics, Chie Mihara produces heels that are well known for their wearability . . . as in all day!

Which fashion footwear brand has evolved and adapted well to consumer demands?

One of my more popular brands is Arche; latex soles and soft uppers make them a very comfortable option, brightly colored and understated.

Which fashion footwear retailer do you admire the most and why?

The British Canadian, London-based designer and retailer Tracey Neuls has my admiration . . . I love her designs, again high in the comfort and wearability zone, her footwear is elegant, funky, and quirky. She stands out as artistic and design led.

What are your tips for consumer trends for the future, e.g., changes in buying behavior?

As we become more health conscious, with commodities such as petrol never likely to be relatively cheap again, will flat shoes ever go out of fashion?!

What advice would you give someone setting up their own footwear store?

Advice for someone opening their own store would be to do your local research thoroughly. I got out on the street with a set of questions and besides having fun I learnt a lot about what local people wanted and I received some good advice. I talked to local retailers too, most of whom were happy to share knowledge. Tap into all the resources you can. Research every aspect thoroughly.

Will you offer a transactional website, or is your business model based around customers being able to try shoes on?

Shoes have to be tried on, and feet being so variable from one person to another an online shop will receive a lot of returns, which means a lot of work and expense, so for the time being customers have to come to the shop.

Which footwear retailer or brand do you think has the best retail presentation and why?

This has to be Tracey Neuls; her shops are like galleries, they are interesting, intriguing, witty, and stylish just like her shoes . . . different, so straight away you know you're going to get something different, and if you like all of the above you can't help but be drawn in.

4.17

4.17 mo Brog, Brighton, UK
mo Brog's main footwear wall features brands such as Audley, Tracey Neuls, Arche, Chie Mihara, Lilimill, and Angulus.

SUMMARY

To understand how fashion footwear is distributed across various retail channels and market levels, understanding the difference between branded and own-label product is key. Retailers need to consider the most effective and profitable means of not only protecting their market share within their domestic market, but also how they can expand into new markets, and which modes of market entry they will select from a retail perspective; they may need to employ different strategies for different markets that may be a combination of retail business models. Due to the complex and varied nature of how these models interact, combined with increasingly savvy and selective consumers, new retail models are evolving that challenge these traditional established models—namely "inventory-less" models and "media retail." Retailers will always be focused on creating and maintaining profitable space, and with these pioneering stores taking innovative approaches to space planning, inventory management, and sponsorship, new methods are being piloted that could have a significant and long-term effect on the retail landscape. However, retailers that are still focused on service and offer an enjoyable and even luxurious shopping experience will continue to be relevant in the evolving retail landscape.

DISCUSSION QUESTIONS

1. Who are the top twenty fashion footwear retailers in your region?

2. What trends can you establish with regard to the increase/decrease in market share of specialty versus nonspeciality stores?

3. How would you describe the retail landscape of the market leader in your region?

4. What are the challenges faced by pure play retailers?

5. What can fashion footwear retailers do to improve the consumer experience in store?

EXERCISES

- Following the Rachel Shechtman "media retail" model, how would you create a footwear version of the STORY concept in your local market? By conducting primary and secondary research, devise a response geared solely around fashion footwear to the seven principles outlined in the case study.

- Select a retailer in your local region that is losing market share and that is currently a multi-channel retailer and has yet to adopt an omni-channel approach. What recommendations can you make with regard to both the in-store retail environment and how this environment is linked to its digital marketing?

- Conduct a focus group to ascertain consumer preferences with regard to "inventory-less" models (as outlined in the Sneakerboy case study). Would this model work for all product types? What are the advantages and disadvantages of this model?

KEY TERMS

Branded goods—product identifiable as being from a particular company and that can be sold via retail and wholesale channels

Concession—where a brand or designer operates an area within a larger store or retail area, most usually a department store

Department store—large retail store with an extensive assortment and range of goods, organized into separate departments

e-tail—retail conducted via the internet

Franchise—when a brand/retailer gives permission to another business entity to sell its product on its behalf in another geographical territory

Independent retailer—small stores with a single location or up to three locations

Inventory-less retail—where product is not held on the retail premises as stock (inventory)

Joint venture—a contractual business agreement between two or more parties

Licensing—a business agreement usually bound by contracts for a fixed term whereby the brand (intellectual property owner or licensor) gives another party permission to design, manufacture, and sell product bearing the brand's name (as a registered trademark)

M-commerce (mobile commerce)—the buying and selling of goods and services through wireless handheld devices such as cellular telephone and personal digital assistants (PDAs)

Market share—total sales in a market captured by a brand, product, or company, expressed as a percentage

Multichannel retail—retailers that sell via more than one channel

Multiple retailer—a retail shop that has multiple locations but has common ownership and management

Private label—a brand owned not by a manufacturer or producer but by a retailer or supplier who gets goods made by a manufacturer under its own label

Pure play—a company that operates only on the internet

Omni-channel—retailers that sell across multiple channels and are perceived by customers to offer a seamless, integrated shopping experience

05

MANAGEMENT STRATEGIES FOR RETAIL GROWTH

Learning Objectives

- Define retail strategy and explain the approaches needed to develop a strong plan for growth.
- Identify the key audit tools used to analyze the external and internal environment in strategic development.
- Evaluate the various financial, pricing, and product development strategies used by footwear retailers to create competitive advantage.
- Assess the impact and importance of strategic action by retailers through internationalization, acquisitions, and brand development in retail.

5.1 Kurt Geiger storefront
Kurt Geiger is currently one of the UK's most successful footwear retailers; through a mix of distribution strategies via concessions, licences, own stores, and multichannel, it offers well-known brands and its own label trend-led fashion footwear.

INTRODUCTION

Retailing footwear is a diverse activity that encompasses a variety of elements in addition to selling tangible products. Fashion styles and trends move quickly, and there are fluctuations in consumers' preferences. The technology to design and deliver product to stores is constantly evolving, and the means of display, communication, and promotion of footwear is now highly sophisticated. In recent years there has been rapid development of online retailing. Previously seen as an option or strategic choice for brick-and-mortar retailers, online is now essential for survival.

As outlined in the previous chapter, fashion footwear retailing is unique and differs in several areas from that of apparel. The strategic direction of a company and how it defines and develops this is crucial for both survival and growth in the market. An understanding of the fast-moving and evolving retail landscape and its environs will help a company to identify its place and set its goals. This chapter will define and examine retail marketing strategies required for growth in the fashion footwear sector. It also will explain the auditing processes of the external and internal environment and how a company uses this information to form a growth strategy.

WHAT IS RETAIL STRATEGY?

A strategy is a plan put in place detailing how a company will deliver its product to the end consumer. The business must understand the external environment that it is operating in, its competition, and the consumer that it wishes to target. It needs a solid understanding of its own strengths and weaknesses, as ultimately the aim is to beat the competition. Plans can be developed from past retail operations and sales performance.

Successful retailers must be able to meet the following requirements: 1) Deliver the goods to the store and customer in a timely way, in the best physical location and/or appropriate online format. 2) Meet customer demand and offer a suitable level of customer service with accurately priced product, which necessitates ensuring that the retailers have appropriate levels of inventory so the customer can be satisfied, but not too much stock so that money is tied up in product that is not selling. 3) Correctly display merchandise that is fitting for the target market and communicate a relevant and appealing brand message.

Retailers use a variety of strategies to develop and grow their business. They may expand into new markets and distribution channels, improve and create additions to their current product ranges, build a differentiated brand image, and develop profitable product designs. Many of these strategies can be achieved through either being the cheapest in the market or being unique among their competitors.

The process of strategic growth can be considered in three key stages:

1. Development: intelligence gathering and outlining the current position of the company both internally and externally.
2. Planning and implementation: defining and establishing a competitive advantage and assessing where the company has potential considering its core competencies.
3. Action and evaluation: using the right strategy in the right location and appraising its success or failure.

STRATEGIC DEVELOPMENT

The first stage of strategic growth identifies where the company is currently placed in the market. Several techniques can be used to survey the market, also known as an audit of the micro and macro environment.

What Are the PEST Factors Affecting the Traditional Footwear Retailer?

POLITICAL
Footwear import and export duties, quotas, and trade agreements (see Chapter 3) are a political strategy used by governments to leverage trade and growth between countries. It affects the price a retailer must charge for goods.

ECONOMIC
Variants in sales tax and from region to region (country and by state) combined with unpredictable exchange rates make price parity challenging for wholesale brands and designers alike, and this pressure is passed to the retailer. Shrewd retailers, and specifically retailers that work across regions, can use this to their advantage; however, tracking and monitoring this is both challenging and time consuming.

SOCIOCULTURAL
Consumers are increasingly time poor, while more people undertake more than one job to make ends meet. e-tail gains from this, whereas brick-and-mortar retail suffers as a result of having to provide longer opening hours.

TECHNOLOGICAL
The rise of m-commerce and apps allows the customer to use retailers as showrooms to browse and try on but then search for the best price and greater choice online (see Chapter 4). Innovations in 3-D printing techniques to create footwear prototypes and custom-fit shoes (see Chapter 2) may also offer better customer service experience and address "fit" issues.

Macro Environment

The audit tools that footwear businesses use do not differ from apparel, but many of the key issues in the external environment do. When conducting a scan of the macro environment, companies will look at the external "forces" that are outside of the company's control. It highlights current issues in the wider marketplace. These are political, economic, social, and technological (PEST) issues, as well as increasing concerns for the environment and legalities within the marketplace.

Micro Environment

The micro environment considers the forces that have a direct link to the company, such as the availability of raw materials, suppliers, and competitors whose actions will have an impact on store management, operations, and expansion of the company.

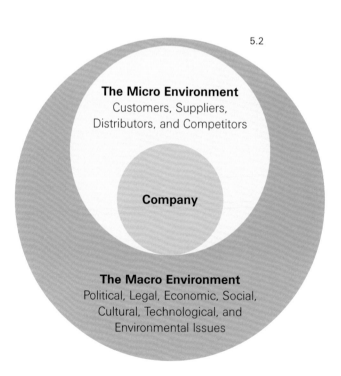

5.2

The Micro Environment
Customers, Suppliers, Distributors, and Competitors

Company

The Macro Environment
Political, Legal, Economic, Social, Cultural, Technological, and Environmental Issues

5.2 The marketing environment
The marketing environment: the first stage of planning involves understanding where the company sits in the wider environment by assessing the micro and macro situations that surround and affect it.

Porter's Five Forces

Developed by Porter (2004) this process measures the suitability or appeal of a sector such as the footwear industry by looking at five competitive forces (see Figure 5.3). The advent and growth of e-tail has been challenging for the independent footwear specialist at local level. High streets and malls are increasingly homogenized, and the buying power of the larger e-tail groups such as Amazon and Zappos are adding more pressure onto independents with less buying power.

Internal Company Environment

Online retailing and global manufacturing means that businesses must be equipped to operate 24/7 and 365 days a year. Therefore, internal resources, capabilities, and weaknesses must be recognized along with an identification of the key leaders and influencers in the company. Assets such as HR, logistics, and know-how, e.g., stock systems, market intelligence, and stores and their location, as well as intangible ones like the retailer's brand name and its market position, must be considered. Increasingly considered are a company's values, specifically those relating to ethical standards and how these are implemented and managed. American e-tailer Zappos.com highlights its core values on its website.

5.3

Threat of substitutes	Can customers go elsewhere, go without, or buy a different product instead?
Threat of new entrants	Are there new retailers and brands coming into the market?
	How easy is it for new products to enter the market?
	Can online retailers access your existing market without opening a store?
Power of consumer buyers and retail buyers	The consumer: How reliant are you on existing customers?
	How do you retain them and attract more?
	Can you meet their demands?
	The retail buyer: What is your power as a buying organization in terms of size of orders, economies of scale, store kudos, and brand awareness?
Power of suppliers	How many suppliers do you have?
	Are they branded or own-label suppliers, and what is your relationship with each one?
	How much does the business depend on them to deliver the right product at the right price?
	If the relationship changed, what effect would this have on your business?
Rivalry among existing competition	Who are your existing competition and what are they doing?
	Do they have new management structures or strategies in place?
	Is the industry growing, and are your competitors profitable, stable, or closing down?

Zappos Family Core Values

As we grow as a company, it has become more and more important to explicitly define the Zappos core values from which we develop our culture, our brand, and our business strategies. These are the ten core values that we live by …

1. Deliver WOW Through Service
2. Embrace and Drive Change
3. Create Fun and a Little Weirdness
4. Be Adventurous, Creative, and Open-Minded
5. Pursue Growth and Learning
6. Build Open and Honest Relationships with Communication
7. Build a Positive Team and Family Spirit
8. Do More with Less
9. Be Passionate and Determined
10. Be Humble

5.3 Porter's five forces checklist for footwear retailers
Porter's five forces highlight key areas in the marketing environment that a company should consider; each "force" should prompt questions for footwear retailers to help develop a strategy in line with the company's vision.

5.4

SWOT Analysis

A SWOT (strengths, weaknesses, opportunities, and threats) analysis requires a candid look at a company's strengths and weakness and how these can be used to develop opportunities and combat threats in the market. A set of questions is used to identify the issues.

QUESTIONS FOR THE SWOT AUDIT

Identifying strengths and weaknesses
• Product: Is it high quality? Is it fashionable? Is it unique? Is there width and depth of assortment? Are they own-/ private-label or branded goods?
• Promotion: How well is the brand known? What marketing and promotional strategy is in place?
• Place: How many retail outlets are there? Are they online and off-line? Where is the product available?
• Price: Who buys the shoes and how much do they cost?
• Personnel: Is there staff expertise in management, buying, and customer service?

Identifying opportunities and threats
• Product: Fashion styles are constantly evolving, and therefore product innovation, development, and diversification are important.
• Promotion: Consumer brand awareness and press contacts and relationships need to be developed and expanded.
• Place: Footwear retail is a highly competitive and saturated market. Is a niche or new location available?
• Price: Can the price be changed/is it profitable?
• Personnel: Employees are brand ambassadors and often a company's most expensive resource; their skills, experience, and behavior can directly affect the success or demise of a retail store.

5.4 adidas Originals
Pharrell Williams collaborated with adidas Originals to create a vibrant collection of shoes, tees, and jackets.

5.5

Strengths	Weaknesses
Strong brand heritage with iconic products that aligns with casual and sports fashion trends.	Dependence on external manufacturers and outsourcing production to contract manufacturers mainly in Asia.
Ability to develop and maintain collaborations with key influencers and celebrities such as Pharrell Williams and Kate Moss.	Susceptible to disruptions in supply chain and quality and price fluctuations due to fluctuating exchange rates and trade tariffs.
Established multichannel approach, across a variety of distribution channels, own retail, wholesale, and online.	

Opportunities	Threats
A consistently growing global footwear market.	Intense competition across all sectors in the footwear market from both private-label and brands, particularly key competitor Nike.
Growing online retail channel as consumers become more confident in purchasing online.	UK's exit from the European Union may affect sales and trade.
Increased innovations in sole and comfort technology in trainer footwear.	Rise in labor wages particularly in Asia may increase the cost of products
Continued casualization of fashion footwear, revivalism, and vintage sportswear trends.	Counterfeit goods market has increased significantly (see Chapter 7), which has the potential to harm the brand.

5.5 SWOT analysis
A SWOT analysis showing the strengths and weaknesses of adidas Originals.

5.6 and 5.7 The Natural Shoe Store
The Natural Shoe Store in London's Covent Garden was one of the first retailers to establish itself in the area over thirty years ago. Its vision to re-create natural walking within the confines of a shoe continues to drive the company to provide high-quality, fashionable footwear that feels great to wear. Its continued success is based on a passion to offer brands and styles that are consistent with the retailer's values: comfort, style, tradition, and craftsmanship.

STRATEGIC PLANNING AND IMPLEMENTATION

The second stage of strategic growth devises profitable ways to expand a business, implement strategy, and satisfy the consumer. A company may also define its corporate mission or statement, which enables a focus on success.

Financial Management and Leadership

Strategic choice is defined by the financial situation of the company and the leadership style and focus of senior management. The fashion and footwear industries are passionate and creative, fueled by designer start-ups and retail entrepreneurs. Many small firms will lack access to large commercial and financial backing despite having a strong personality at the helm. Retail businesses are reliant on consumer spending to maintain operations, so it may be difficult even for an existing medium-sized retailer to raise capital. This leaves the business vulnerable to stagnation or even takeover. For large corporations it is often the strategic management team and shareholder expectations that will define the plan for the company.

Raising Capital

Funding for development and expansion may come from a variety of sources. Traditional routes to raise finance for large corporations are through initial public offering (IPO), which can raise millions through the sale of company shares to the general public. In 2014 Jimmy Choo raised 140 pence per share in its IPO on the London Stock Exchange. The company was valued at £545.6 million ($877 million), according to CNBC.

Recent trends have seen the use of online platforms to raise capital, known as crowdfunding or crowdsourcing. Websites such as Indiegogo, Kickstarter, and Crowdfunder allow a start-up company to promote its business in a bid to raise a set amount of money to invest in a new product or next season's range. Funders pledge an amount of money and in return receive a thank-you, acknowledgement of support, and in some case a tangible return such as a pair of shoes.

The effect for the company can be twofold; while raising capital for a project or product, it can also gain validation for its concept, raise awareness, and connect with customers. Creating evangelists is increasingly important in the era of online social endorsement and selling. Indiegogo suggests that between 31 percent and 75 percent of crowdfunding backers are friends of friends (those who have pledged and shared online). Companies such as the Boston Boot Co. and BANGS Shoes in Brooklyn, New York, have successfully met their financial goals and raised their profile at the same time.

5.6

5.7

Price Architecture and Strategy

The price of a product is a direct determinant of profits or losses for a company. It determines the type of customer who shops in the store and the other brands and retailers it competes with; therefore it affects consumer perceptions of the brand's image, quality, and prestige. Companies need to benchmark their costs against their competitors' costs to learn whether they are operating at a cost advantage or disadvantage. Awareness of the price and quality of competitors' offers will determine if the company's offer is superior, in which case it may charge more than the competitor.

When setting prices it is important to make sure that a range of products are available at each of the key pricing bands. This is known as price architecture, or a structure that encompasses the lowest to the highest prices that the brand or retailer needs to maintain sales and market share and remain profitable. Once costs and margin have been established (as discussed in Chapter 3), key price points will be worked into the price architecture and strategy.

Creating Competitive Advantage

Creating an advantage over the competition is achieved in two ways: by being the cheapest or the most different in the market. Customers ultimately put a value on the product by showing how much they are prepared to pay for it.

1. Low-cost leadership: retailing a popular, generic or unbranded shoe, such as a flip-flop or pair of slippers, to a large market share at the lowest possible price allows the retailer to become the market leader because it is selling the highest volume of product.

2. Differentiation: retailing a product that is unique, exceptional, or exclusive and so the retailer can charge a premium price for it. In fashion footwear, differentiation is usually developed through brand identity (explored further in Chapters 7, 8, and 9), which also allows the retailer to charge a premium price.

The pricing strategy that a brand or retailer chooses will depend on which type of competitive advantage it uses. While the cost of sourcing must be considered (see Chapter 3), the final price must consider the brand's quality and promotional strategies relative to competition.

5.8 Price architecture
Retailers set prices that are both appealing and accessible to the customer as well as maximize profits for the company.

5.8

High price
Usually stocked in key flagship stores or limited distribution; also maintains exclusivity and aspiration for the brand or product.

Premium price (bridging products)
Depending on brand position and perception, these can be either the top end for high street retailers or the lower price point for designer brands, allowing access to the middle market.

Mid-price
Volume product that consists of best-sellers and core items for mid-level and mass-market retailers.

Lowest price
Entry-level products, small accessories, shoe care products, and laces to increase brand awareness or cheap-to-produce footwear with seasonal appeal, such as flip-flops.

5.9

5.9 Low-cost leadership at Matalan
Fast fashion clothing and footwear retailer Matalan offers a huge variety of simply designed ballet flats in a variety of colors at competitive prices. This satisfies demand for cheap, fast fashion footwear.

Pricing strategies	Description	Example
Premium pricing	A high price is used where there is a uniqueness about the product, where a substantial competitive advantage exists. Relies on high brand awareness and desirable product.	Designers such as Charlotte Olympia and Manolo Blahnik and premium department stores like Bergdorf Goodman and Harvey Nichols.
Penetration pricing	The price charged for products and services is set artificially low in order to gain market share. Once this is achieved, the price is increased.	Clothing retailers may initially enter the footwear market with shoes that are priced to make them attractive to their existing customers to make them switch from shopping elsewhere, e.g., New Look.
Economy pricing	This is a no-frills low price. The cost of marketing and manufacture are kept at a minimum, and brand awareness is not relevant.	Supermarkets such as Walmart and Asda selling own-label footwear sourced directly from manufacturers.
Price skimming	A high price is charged because of a substantial competitive advantage. However, the advantage is not sustainable. The high price tends to attract new competitors into the market, and the price inevitably falls or is heavily promoted due to increased supply.	Linked to fad products that are initially very appealing to the fashion market, so they can command a higher price, such as Crocs. However, styles can be easily replicated, and inevitably the price is reduced.
Psychological pricing	This approach is used when the marketer wants the consumer to respond on an emotional, rather than rational, basis.	Brands such as Gandys, whose social mission is to raise funds for orphans through the sale of its flip-flops, can rely on an emotional or "purchase with purpose" response from customers.
Promotional pricing	The original price is lowered for a set amount of time and is used to promote a product.	Discount Shoe Warehouse's (DSW) reputation is built on finding a bargain and uses "price" reductions as the reason to solicit a sale.
Geographical pricing	Geographical pricing is evident where there are variations in price in different parts of the world.	Variations may be due to duties and logistics costs and are more transparent than ever due to the internet; therefore, every brand must make this seamless for the consumer.

STRATEGIC ACTION AND EVALUATION

The third stage of strategic planning is using the right strategy in the right location and appraising its success or failure. Chapter 4 identified seven key types of retailers of footwear (see Figure 4.2) based on product mix and store format. Each retailer has the opportunity to move into a new market, develop its retail format, or extend its product selection. Ansoff's (1957) growth matrix identified four main strategies that retailers use to grow their market share.

Own Brand/Private Label Development

Branding can provide a point of differentiation from the competition; it identifies the product or service of the retailer and makes it distinct from anyone else.

Developing an "own brand" or "private label" gives the retailer a competitive advantage over its rivals in a highly sophisticated and busy market. The customer must visit that store or online portal to buy those shoes. It gives the company control over product design, packaging, and price, as well as the merchandise selection, stock levels, store, and online environment. Successful recent examples of this strategy are multiple clothing retailer New Look, which is now the second-largest retailer of value fast fashion women's footwear in the UK.

5.10 Ansoff's growth matrix
Retailers will go through a process of one or more of the four growth strategies outlined by Ansoff to implement their vision and expand.

5.10

Market penetration; offering existing products in the existing market

Maximizes return of financial investment to create profit.

The fashion industry is based on obsolescence: consumers expect new designs, at the very least on a seasonal basis. To maintain market share, retailers must deliver a consistent service and product. To grow, they must improve productivity.

Gaining a reduction in the cost price paid for a product, i.e., getting it cheaper from the supplier but not passing the saving onto the consumer.

Increasing the brand reputation and appeal of the retailer's own private label.

New products; offering new products to the existing market

Extending ranges to offer menswear in addition to their existing women's line, or add a collection of sandals in summer or boots in winter.

If they have a strong brand name, they may also develop an entirely different product category such as clothing, sunglasses, or perfume.

A clothing retailer may develop a footwear range using this strategy. This is known as product diversification through retail own brand/private label.

Market development; offering existing products in to new markets

Store expansion—opening new stores as well as creating a transactional website. Internationalization—expanding the existing brand and retail operations to another country (see text that follows).

Expansion; offering new products to new markets

This uses a mix of market and product development, and although it is the riskiest of all four strategies, it can reap the highest returns.

However, problems may arise; there is a heavy commitment to design and product development and a huge cost commitment to produce for multiple stores. The risk of late deliveries and poor customer response to new designs may put the whole company in jeopardy. Footwear manufacturing, as discussed in Chapter 2, is a complicated and costly activity with many differences from clothing production. Consequently, many clothing retailers have traditionally turned away from extending their product ranges to include footwear. However, increased communication, global production, and availability of product have allowed nonspecialists access to suppliers in Asia that were either not equipped or not in existence twenty years ago. Footwear as a new product category can potentially offer higher margins than other products, leverage over the manufacturer, and increased loyalty and purchasing from existing and new customers.

Internationalization Strategy

Increasingly globalized and homogenized consumer tastes mean that as existing markets mature, new, foreign markets are becoming attractive to retailers. Opening a store in a new country offers an element of prestige and highlights to the competition a retailer's position in a new market. For luxury brands, it may spread the risk or prevent overreliance on one or two markets, such as the United States and Europe, and in a protracted economic downturn they may look to Asia for further growth and sales.

Where to internationalize may be decided by countries that have a close geographic proximity or share similar language, psychological, or cultural traits, e.g., United States to Canada or UK to Ireland. Counties with favorable political and trade policies—for example, those that are members of the European Union—are also attractive when looking to enter new markets. There may also be a more ambitious motivation to tap into larger, increasingly prosperous markets like China or under-developed markets with potential, such as India.

5.11

5.11 Top Shop diversification into footwear
Owned by the Arcadia group, Top Shop is an innovative fast fashion retailer with a well-established and large market share in clothing. Over the last decade it has diversified to include its own-brand/private-label footwear, and it is now a key product category in the store.

The PLIN (product, lifestyle, image, niche) model presented by Akehurst and Alexander (1996) suggests that success for specialist retailers, such as those selling footwear, is dependent on four key areas.

P—Product: What kind of product is it? Is it different from others on the market?

L—Lifestyle: What are the customer's shopping habits?

I—Image: What kind of image does the brand/retailer have?

N—Niche: Is there a space in the new market for this brand?

How to internationalize will depend largely on access to skills, finance, and the strategic plan, as well as opportunity. New markets have different competitors and customers; therefore, retailers will need to either adapt their products and operations to suit or take a global approach and keep consistency. Different marketing strategies for this are discussed further in Chapter 7.

Due to the niche nature of footwear products, internationalization entry methods are often through internal expansion, franchising, and joint ventures (discussed in Chapter 4 in modes of market entry). Internal expansion or organic growth requires large financial investment and risk, but it is often the next or logical move for higher-priced brands as they can maintain control of the brand image and distribution (see case study at the end of this chapter).

Acquisitions

For financially stable companies, choosing to acquire a less well-known or struggling brand can add strength and diversity to their portfolio. It allows them entry into a market sector such as footwear that might not have been possible. Acquisitions may happen when a market is well established and mature, meaning there is little time and opportunity to grow organically. Acquisitions also give opportunity to take out a competitor and gain some source of competitive advantage, as well as the option of picking up a bargain during a tough economic period. The advantages for the company that has made the acquisition means that it is gaining assets in the form of product designs, corporate and intellectual property, favorable leases and locations, access to distribution channels, an income stream, knowledge, and skilled/key people. These are all sources of competitive advantage that will allow the company greater power in the new marketplace. For the acquired, it holds the advantage of greater investment and expertise from a parent company and for those struggling, a reprieve. Recent examples of acquisitions of footwear retailers include Zappos.com bought by Amazon in 2009, and in 2012 shoe e-tailer Cloggs was bought out of administration by sportswear retailer JD Sports.

Evaluation

No retailer can afford to stand still; fashion is fast moving, and regardless of retailer type, companies must be aware of their place in the market and of their customers. The nature of fashion footwear allows it to be sold in different store environments, both specialist and multiproduct; however, its success will be attributed to its ability to sell as a product in its own right by creating a profit for the retailer. Once the strategic plan is in place, there are several ways of evaluating its success, but ultimately it comes down to two factors: are customers buying the product, and is it making a profit?

" HOWEVER BEAUTIFUL THE STRATEGY, YOU SHOULD OCCASIONALLY LOOK AT THE RESULTS."
WINSTON CHURCHILL

ETHICS IN ACTION:
International Cultural Considerations

When entering a new market or sourcing product from a foreign supplier, consideration must be given not only to the culture of a country but also to the market and business culture in the host country. Hofstede (1984) originally identified four dimensions of cultural considerations, which he later developed to include six, that offer insight into the behaviors different cultures exhibit in the workplace. While this text makes generalizations, in an increasingly globalized marketplace we need to be mindful about how our actions may be interpreted by other people and gain a deeper understanding of how workers behave when presented with ambiguity, risk, and inequality, both as individuals and as a group.

5.12

Individualist versus collectivist society

Western, developed countries such as the United States display traits of self-interest with loose ties or obligations to wider society and a focus on individual attainment, personal goals, and desire for acknowledgment. However, countries such as Bangladesh and Mexico are considered a collectivist society, with individuals exhibiting loyalty and interdependence in the group. This is critical to overall satisfaction and achievement in their life.

Uncertainty avoidance—coping with ambiguous situations

Behavior in a society can also be measured by how workers respond to the enforcement of rules. In many of the Latin American countries, such as Brazil, workers want to avoid uncertainty or ambiguity and need clear instructions from a line of authority. Rebellion is not tolerated. Whereas the Danish work environment is more relaxed and flexible; rules and guides exist but there is a high toleration of risk and less reliance on leaders to rule.

Large versus small power distance—inequality

A large power distance culture such as Russia has an unequal distribution of traditional notions of power and wealth. It develops dependence by subordination through authority and elements of protectionism. However a society where everyone is considered equal, such as Australia, has a small power distance culture and creates interdependence and trust.

Masculine versus feminine

Chinese business practices often exhibit traditional "male" values: achievement, power, and materialism. "Female" values, including solidarity, equity, and concern for others, are seen as important in Vietnam. While many countries have a legal framework that does not allow differentiation of jobs based on gender, these soft dimensions need to be considered.

5.12 Aldo storefront
Canadian brand Aldo first established in the US market in 1993 and now has over 1,750 stores worldwide.

> **" CULTURE IS LEARNED AND SHARED AMONG A GROUP OF PEOPLE. PEOPLE FROM A GIVEN CULTURE EXHIBIT SIMILAR NORMS AND VALUES AND THEREFORE, CULTURE AFFECTS THE VALUES FOUND IN THE WORKPLACE."**
> MEHTA ET AL. (2010)

CASE STUDY:
TANYA HEATH Paris—International Innovation in Fashion Footwear

TANYA HEATH Paris is a global luxury brand combining fashion and footwear innovation to create transformative footwear for the modern woman. It is the first company in the world to create and sell a multi-height shoe with workable removable heels, allowing the wearer to go from 4 centimeters to 9 centimeters on the same shoe. There are several styles of heels that come in a huge variety of colors and finishes. As well as interchangeable heels, the range also offers different shoe styles—pumps, sandals, low-boots, and boots.

The brand was established in 2012 by Tanya Heath who, after studying international relations in Toronto, moved to Paris in the early 1990s. In parallel to her career in management consulting, high-tech start-ups, and private equity, she lectured in disruptive innovation for eight years at one of France's leading engineering schools. All of this gave her the grounding to explore the idea of an innovative bespoke shoe that could change in height depending on the wearer's preference.

Product Development

She took a unique approach when developing the concept of the detachable heel, deciding not to take the typical research and development methods for fashion and footwear designers and worked with mechanical engineers in France to create the patented detachable heel and shoe. It took fourteen engineers two-and-a-half years to perfect the system that embeds the mechanical heel clip. Over ten designers, shoe technicians, and last specialists worked on creating styles that could adapt to a multi-height.

Made in France

All of the shoes are designed and made in France, allowing for high-quality and small production runs. This flexibility means that the team can adapt quickly to trends in colors and work closely with the factories and suppliers to respond to consumer demand. Tanya is keen to source components and leathers within Europe, where there is greater ethical and environmental transparency in the supply chain. The majority of materials are from France, with a few additions from Portugal and Italy where necessary.

5.13 TANYA HEATH Paris
Tanya Heath Paris is the first company in the world to create and sell a multi-height shoe with interchangeable heels.

5.13

5.14

Internationalization

As the first company in the world that sells not just uniquely engineered shoes, but also detachable heels, TANYA HEATH had challenges that no other shoe company had to consider. It addressed questions such as: do the shoes need a user guide, would people be willing to buy separate heels, and how do you merchandise and sell extra heels? It also had to consider how it would expand abroad. What partnerships needed to be developed, and could the shoes be wholesaled to other retailers or would it be better to open TANYA HEATH Paris stores?

After spending several years getting the product right and identifying the type of woman who would appreciate the concept and wear the shoes, all of the international retail expansion has been organic. As a luxury brand, it is in its infancy, and relationships with new business partners have evolved from those who have discovered and worn the shoes or love the stores and want to be part of the evolution of the unique concept.

TANYA HEATH stores are now in Europe: Paris, Porto, Lisbon, Lausanne, Luanda (Angola), and Amsterdam as well as in North America: Toronto and Los Angeles. The more established stores offer custom heel making at a heel bar, where a woman can choose the finish of her heel. But all stores feature heel walls and the same product and the same knowledge and service. Currently 35 percent of sales are from the flagship Paris store, but the brand is also seeing the increasing importance of its website http://www.tanyaheath.com, which offers a seamless experience for its growing global customer base.

5.14 Innovation and comfort in one shoe
The TANYA HEATH Paris collection is available in a variety of heel heights and shoe designs.

CASE STUDY:
TANYA HEATH Paris—International Innovation in Fashion Footwear

Wholesale distribution into multiband stores is complex. The product is a new and innovative concept and a brand that requires a high element of customer service and education about how the innovation works. The purchase experience in store is vitally important, not least because the shoes are around 300 euros ($350) but also because the heel options and colors are vast, so guidance from a trained sales consultant is essential. This experience generates considerable positive "word of mouth" promotion that could be lost in a large multiband environment.

There are the inevitable challenges and barriers to growth that affect most footwear companies, particularly high tariff barriers (35 percent or 40 percent in Brazil when exporting from Europe). Many new brands have to monitor their margins very carefully, and TANYA HEATH is no different. Some markets such as Brazil are unapproachable until there is greater brand awareness or a potential to licence production in the local country; however, this may bring different challenges for the brand. Constant geopolitical change can also make investment decisions for the company difficult, such as the UK's recent decision to exit the European Union's free trade zone, making a store in London less straightforward to achieve.

Success

Tanya is an innovator, and the brand has won many awards and accolades since its introduction. She has embedded in the brand's DNA the story of problem solving—suffering in uncomfortable shoes is a problem that affects countless other women, but by doing things differently and by using innovation and technology to create a new type of shoe, this resonates across nationalities and age groups. Fostering loyalty is key, so much so that in the Paris store over 50 percent of customers have returned to buy a second pair of shoes.

> **" OUR CUSTOMER LOYALTY IS JUST PHENOMENAL AND IT MAKES ME BELIEVE THAT OUR PRODUCT WILL FIND ITS PLACE AND THAT THE BRAND HAS A STRONG FUTURE."**
> TANYA HEATH

5.15

5.15 TANYA HEATH store in Toronto, Canada
Interior view of the store in Yorville, Toronto, which includes the iconic "heel wall" that adds to the interactive consumer experience.

Industry Perspective:
Marc Goodman, Managing Director, Giancarlo Ricci, UK

Menswear store Giancarlo Ricci has been trading for the past twenty-eight years and is Liverpool's premier destination for designer fashion. The family-owned business is now run by Marc Goodman, whose innovative and ambitious growth plans in the past few years have included pop-up stalls at sneaker fairs and opening a men's footwear store.

What are your strengths and weaknesses as an independent footwear retailer?

Giancarlo Ricci is an independent clothing store that opened a footwear store in 2012. We are unable to buy volume, and every percentage of margin counts, but we are able to identify what people are wearing on the street and get it into store fast. We are responsive to trends and offer our customers a selection of products from the world's leading luxury brands in a stylish and friendly environment.

What opportunities and threats can you identify in the market today?

The big sportswear retailers are acquiring smaller menswear independents in the UK, and our business is about keeping brands precious. Unfortunately, they discount aggressively and dump product when it doesn't work—this can be a major threat to our reputation and profit margin. Brands have equity and they have to control their distribution and not allow themselves to be milked by the larger buying groups. Independent retailers who maintain a city center location will have the edge so long as their product remains specific to the city and market they serve.

Why did you choose to offer new product (footwear) to an existing market (clothing consumers)?

Brands like Hugo Boss were increasingly offering footwear as a product category, and if it is commercial enough we will buy it. We saw a huge growth in footwear turnover within the main store, so we decided to lease a shop across the road. Footwear seemed the most obvious product to move out as it wouldn't damage existing apparel sales. I didn't have a desire to open a shoe shop, but we needed to expand this category.

We are now spread across two separate stores; some people will not cross the (pedestrian) road to the other store but we do have another frontage—double the window space, with extra staff and extra security. We have increased our apparel offer in the original store and our footwear in the new store. Apparel sales have increased as a result of opening the footwear store, [and] overall profit has gone up.

ARY

Essentially a company must understand the market that it operates in. A variety of techniques are used, such as a PEST analysis and a scan of Porter's five competitive forces. Once this is established, along with knowledge of their consumer, companies must assess their own strengths and weaknesses (SWOT) to identify opportunities and threats to their business. External challenges for footwear retailers in recent years have been the global recession, changes in consumer demands, and sourcing directly from cheaper suppliers in Asia. Competitive forces have changed the landscape for retailers, and the biggest shift has come from clothing retailers, sports brands, and supermarkets now offering fashion footwear. This has led to demise in the traditional shoe shop as a retail format and a growth in private-label footwear sourced directly from factories at a cheaper price.

Retailers must understand their customer and be flexible and open to opportunities as they arise, locally or globally. A strategy is important for all sizes of companies, from the small web-based entrepreneurs to the established high street retailers who are challenged by rising prices and increased competition. Companies may choose to be the cheapest product in their competitive field, the most unique or exclusive, but no two companies will choose the same strategy.

DISCUSSION QUESTIONS

1. Assess the main pricing strategies and find example retailers in your region.

2. What factors might lead a company to a) acquire another one or b) allow itself to be acquired? Give examples.

3. Outline the key differences between a low-cost leadership strategy and a differentiation strategy. Give examples.

4. Conduct a SWOT analysis on a footwear brand or retailer of your choice.

5. What are the key issues for independent footwear retailers? What strategy can they develop to increase market share?

EXERCISES

1. Choose a footwear retailer that you are unfamiliar with and conduct a marketing audit and assess it from three different angles, considering the following:
   a. Where is the company now in the marketplace?
   b. Where do you think it should be?
   c. How could it achieve this strategy?

2. In your market can you identify the four key typologies of retailers that sell footwear? In each case, give an explanation and example to justify your choice.

3. Do you believe a national culture exists? Use the cultural comparison tool to assess your country's cultural dimensions (http://geert-hofstede.com/countries.html). Choose a country that your country trades with and assess the differences and consider how these may be overcome.

KEY TERMS

Administration—a company is unable to carry on trading due to financial instability and is managed by an external agency on behalf of the creditors

Ansoff's growth matrix—a model to determine the growth strategy for a company

Acquisition—the purchase or takeover of a company by another for purposes of market growth

Core competencies—the central strength a company may base its growth strategy around

Crowdsourcing/crowdfunding—generating knowledge, ideas, and funding capital for a business venture through an online platform

Culture—typically a group of people who exhibit similar behavioral characteristics based on where they are from

Differentiation—competing in the market by offering a unique or changed product

Geographic proximity—counties that are grouped close together and may share similar market characteristics

Hedging currency—agreement with a bank to protect the rate of exchange when purchasing goods in different currencies in advance of payment

Homogenized—standardized or regulated offering based on consumers' tastes

Initial public offering (IPO)—shares in a company are offered on sale to the general public through a launch on the stock market

Internationalization—the process of developing a retail business outside of the home market

Low-cost leadership—competing in the market by being the cheapest

Market intelligence reports—reports such as those by Mintel that offer competitor and consumer insights into the markets or sector a company operates in

Mission statement—a summary of the corporate values, aim, and role of a company

Micro environment—the current situation that a company is in and how it operates

Macro environment—the wider context that a company operates in

PEST/PESTEL analysis—an assessment of the external factors that affect a company

Price architecture—structure used to develop the most appropriate pricing policy for a company

Price strategy—the approach used depending on the type of retailer, product, and profit required

Porter's five competitive forces—a model to determine the key competitive issues facing a company

Strategic planning—development of a proposal that considers how a company maintains growth and success

SWOT analysis—an assessment of the internal behavior of a company

06

VISUAL MERCHANDISING AND DESIGN CONCEPTS FOR RETAIL, E-TAIL AND WHOLESALE

Learning Objectives

- Define visual merchandising for fashion footwear and discuss the traditional shoe shop environment, as well as which elements are relevant in today's retail and e-tail landscape.
- Identify what makes for impactful and effective footwear display, via the understanding of key visual merchandising principles.
- Discuss the challenges of visual merchandising for fashion footwear and explore consumer experience strategies such as customization and personalization.
- Discuss the importance of the role of visual merchandising, for wholesale and public relations.

6.1 Vintage and new-season Gucci at Dover Street Market, London
Best-in-class visual merchandising: excellent use of color, props, and lighting at Dover Street Market (London); a multibrand retail concept created by Rei Kawakubo, of Comme des Garcons.

INTRODUCTION

Through the exploration of key visual merchandising and design principles, this chapter discusses some of the challenges that are specific to footwear and how these obstacles are overcome through the use of innovative and practical solutions in wholesale, retail, and now e-tail.

Instead of merely focusing on retail presentation, it is important to consider how retail buyers are enticed into writing orders at trade shows: what visual devices are employed to make them stop in their tracks, just as a customer does on the high street or in a mall? In a crowded marketplace, carving out a point of difference visually needs to begin at wholesale, be established at showrooms and press events, and then be consistent with the brand's or designer's image at retail—a continuous visual narrative that attracts and engages. With the increasing relevance of digital platforms, for both purchases and marketing purposes, the design and functionality of these platforms is now an essential part of the customer's "journey to sale" (JTS).

Footwear presents a fundamentally different set of challenges than apparel or accessories in terms of visual display. Storing full-size runs of each style takes up a significant amount of cubic capacity, far more so than other product categories. With the increasing pressures of real estate costs, operating the traditional footwear speciality model of footwear-only stores (that hold full-size runs of all styles stored in stockrooms off the shop floor) does not make the commercial sense that it once did. New retail models explored in Chapter 4 require different kinds of display techniques that are complementary to these new business models.

WHAT IS VISUAL MERCHANDISING FOR FASHION FOOTWEAR?

An understanding of what visual merchandising is, how it works, and how it can impact upon commercial success is essential not only for visual merchandisers but also for many other roles such as retail designers, buyers, merchandisers, sales associates, brand representatives, and managers. Visual merchandising is a combination of art and science that goes way beyond window displays.

Ultimately, the purpose of visual merchandising is to increase the sales and profitability of specific product categories, in sync with company retail and wholesale strategy, and to persuade consumers and retail buyers to purchase and editors to include styles within their publications.

Visual merchandisers must possess a combination of creative and commercial skills in order to create displays, environments, and content that appeals to existing customers and to attract new customers, in line with company product and retail strategies; they are also brand custodians by ensuring that the image and personality of the brand is accurately reflected at retail, e-tail, and wholesale.

Visual Merchandising at Retail

Retail visual merchandisers (sometimes called retail stylists) are responsible for creating impactful and comfortable environments; window displays, brand adjacencies, percentage of space in store, best-seller and worst-seller analysis, managing inventory levels, use of graphics and "point of sale," color, lighting, temperature, music, fragrance, and interactive elements all fall within the remit of a visual merchandiser. From a footwear perspective, different approaches are required depending on the market level; at luxury and premium levels, only single shoes are displayed, and this necessitates greater levels of service, whereas at mass market and value level a "self-service" model is employed whereby multiple pairs per size are displayed on the shop floor (see New Look store later in this chapter).

6.2 Harvey Nichols shoe department, Edinburgh
Harvey Nichols shoe department, Edinburgh, by
Four-by-Two retail design and branding agency.

Visual merchandisers can either be based in one store
(such single-sited roles tend to be more creative roles
and can involve window design and the production and
maintenance of props) or work across multiple sites and
are responsible for a territory; these roles tend to be more
business focused and analytical than the more creative
single-site roles.

> **"VISUAL MERCHANDISING PUTS THE ART
> AND DESIGN BACK INTO RETAIL. IT PLAYS AN
> IMPORTANT CREATIVE AND COMMERCIAL
> ROLE ... IT IS ESSENTIAL FOR ANYONE
> WORKING IN THE RETAIL INDUSTRY TO FULLY
> ENGAGE WITH THE PRINCIPLE STAGES OF
> DESIGN RESEARCH AND REALIZATION IN
> ORDER TO STRENGTHEN A VISUAL BRAND
> PROPOSITION IN THE MARKETPLACE."**
> SARAH BAILEY AND JONATHAN BAKER (VISUAL MERCHANDISING
> FOR FASHION, 2014)

Visual Merchandising at E-tail

Digital visual merchandisers are still responsible for the
consumers' "journey to sale," but with a greater focus
on technical skills and merchandising analytics. They can
also be responsible for creating and uploading images,
web design, and content creation and optimization; as
these digital roles develop, they call for very different skills
from that of retail specialists in this field. From a footwear
perspective, viewing product from multiple angles and
distances and styled with different outfits is crucial to
creating attractive, functional, and profitable pages. While
these roles usually sit within e-commerce teams, for smaller
brands and non–pureplay retailers that have transactional
sites, these roles can often be "bolted onto" existing roles.
Creating and managing transactional pages, content-only
pages, social media accounts, and blogs can all fall within
the remit of a digital visual merchandiser (this is discussed
in depth in Chapter 9), especially as this role evolves.

SPACE AND SALES

In both cases (retail and e-tail) the customer's JTS is
controlled and space productivity is monitored, and the
percentage of space is adjusted as a result of analyzing
sell-through.

Visual Merchandising at Wholesale

Enticing retail buyers to purchase at trade shows and in showrooms, and selling the idea of a seasonal collection to editors and influencers at press events requires very similar skills to that of retail visual merchandiser. However, the more temporary nature of such shows, installations, and events can allow for a high level of creativity without the security constraints of retail. Driving sales is not the primary objective; instead, creating highly innovative, albeit temporary, brand environments is the priority, so these concepts and schemes can be more experimental.

These visual merchandising specialists are either employed directly by the brand or by an agency; such as Lucky Fox based in Brighton, UK.

Buyers and editors are invited to "experience" a brand, and this is often geared around a seasonal theme, subcategory, or even specific style, such as the adidas Originals re-launch of the Superstar and Stan Smith via a PR and influencer event orchestrated by the John Doe PR agency in London (http://www.johndoehub.com/work).

FEATURES AND BENEFITS

In all three scenarios, designing and creating engaging spaces and displays that communicate the product's best features and benefits are crucial to the function of successful visual merchandising.

THE TRADITIONAL SHOE SHOP ENVIRONMENT TESTED

With staff, service, and storage all being key factors for success for traditional footwear stores, the financial demands of these three precious commodities pose real challenges for today's retailers that wish to follow this long-established model. Retailers such as Russell and Bromley (established over 100 years ago, with forty-three stores in the UK), Clarks, and the Brown Shoe Company still focus on these core priorities. In the digital age, the emphasis that they place on quality, comfort, and fit and how this is communicated through their retail presentation has become integral to their survival. Their reputations are built on upholding these priorities, which were once prerequisites for speciality shoe stores, but in the changing retail landscape are becoming less of a priority for many footwear retailers, or rather, these priorities are being addressed via more cost-effective means (for example "inventory-less" models as explored in Chapter 4 via the Sneakerboy case study).

6.3

6.3 Clarks children's fitting area, Bicester Village outlet, UK
Efficient and functional space with flip-down seating for foot measuring is enhanced by digital screens that convey seasonal themes and educational content.

RETAIL DESIGN FUNDAMENTALS

The term "Atmospherics" is one that Philip Kotler employs in *The Retail Journal* (1973) to describe the main disciplines that are integral to the design of effective commercial spaces. These are: architecture; exterior structure; interior design, and the design of window displays.

Kotler describes this conscious process as "the effort to design buying environments to produce specific emotional effects in the buyer that enhance purchase probability." Kotler has suggested that good retail design can capture and maintain customer interest by encouraging customers to lower their psychological defenses through the stimulation of the four main senses (excluding taste), to provoke a purchase or a return visit to make a purchase. The use of fragrance is a particularly powerful and effective purchase trigger. There is a strong correlation between smell and memory, and some retailers use "premium scenting" as a strategy to help create a positive in-store experience. Recent research also indicates that the fragrance is more effective when it is simple and instantly recognizable, such as citrus, leather, or sea air, rather than a complex and difficult-to-distinguish fragrance.

Music is also a strong influencing factor within the retail environment. There is a correlation between different types of music played and time spent in store, and, interestingly, perceived time spent in store; the key is for the retailer to match music to appeal to the target consumer: this is particularly relevant when customers are queuing to pay for product. If they are enjoying the music in store, they perceive their time waiting as being less and are less likely to lose patience and abandon a purchase.

Retail architects, designers, brand managers, marketers and retail directors work together in the initial design stages to agree on the following store design objectives:

- The location, to be relevant to the brand/designer image and to appeal to existing customers.
- That the ratio of shop floor space and off-sales areas (e.g., stockrooms and staff rooms) enables profitability, given the product type and projected sales forecasts.
- The shop floor design is compatible with the product type, i.e., the right number of fixtures per product category, influenced by the required product assortment (width and depth).
- That materials and fixtures are sympathetic with the existing brand image, or to encourage a shift in perception based on a new strategic direction.
- To move the customers to every area of the store and guide their JTS; the layout will reflect how the existing customer already shops and encourage new buying behaviors, e.g., in new product categories, or subcategories.
- To ensure costs are planned and tracked according to budget (such as, but not restricted to: rental/lease costs, tax, consultancy fees, legal fees, design fees, in-house teams, materials, fixtures, overheads, plant/equipment hire, subcontractors, health and safety provisions and certifications, inventory/stock, and retail staff costs).

These objectives are crucial in the successful design and execution of the initial stages. Further in the process, retail buyers and visual merchandisers will be involved, together with retail marketing and the in-store retail team. The above is a general model that applies to many different retail formats, e.g., department stores, flagships stores, and independents. If the above factors are not agreed upon and addressed in the early stages, this can cause delays and misunderstandings or confusion, which can add costs and disrupt the smooth running and profitability of a project.

How Can Journey to Sale Be Guided?

The JTS is a pathway in-store that the customer is encouraged to follow by the physical placement of entrances, displays, fixtures, signage, elevators, escalators, and exits. The placement of these elements should be part of the initial retail design in order to guide customers around the store by steering them through multiple product areas that were not necessarily their planned destination in-store, and in doing so increasing the chances of more purchases.

Altering these pathways can be achieved by moving free-standing units (FSUs) and "lead in" tables (tables that house product at a lower level, usually toward the front of a store) and also signage.

If we consider three basic types of retail floor plans that have been used in traditional retail formats, we can explore this in the context of fashion footwear.

RACE TRACK LAYOUT

This layout is typically used by department stores to encourage journeying around and through multiple product categories. Best-selling categories and lines will be placed in destination zones, such as feature walls and near the perimeters, meaning that the customer will have to travel through a combination of new or slower-selling lines in order to reach these areas. The direction of travel is dictated by the placement of escalators, which again increases the number of product zones or "mats" housing different brands or product categories.

For example, the shoe hall at Saks Fifth Avenue, New York, is a huge department (that even has its own ZIP code), and its customers are guided around its 17,500 square feet by the strategic placement of luxurious seating as well as fixtures and product.

6.4

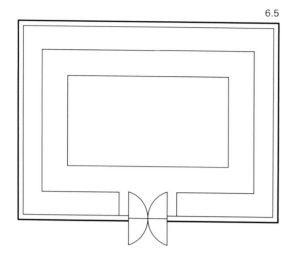

6.5

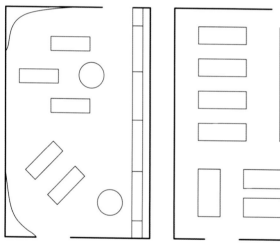

6.6

6.5
From top clockwise; race track, grid, and free-form floor layouts.

6.6 Shoe "maze" at New Look (grid layout)
Profitable use of sales space at New Look, offering the customer plenty of choice and sizes.

GRID

This is a format traditionally used by supermarkets and large pharmacies; however, value and fast fashion retailers such as New Look (Established in the UK with 800 stores globally) utilize this format specifically for their value-driven high-fashion footwear styles. Linear high rise fixtures are arranged in parallel to create a footwear "maze" effect whereby the customer cannot escape without buying!

This is a visual merchandising technique borrowed from FMCG (fast-moving consumer goods) and applied successfully to fashion footwear. This technique is also successful as it maximizes the number of units on the shop floor by displaying full-size runs, negating the need for a large footwear stockroom, which is not a profitable use of space.

FREE FORM

This format is used in smaller retail arenas such as independent boutiques and specialist departments within larger stores; fixtures and aisles are arranged asymmetrically at different heights to create a more organic feel. Larger stores such as Anthropologie and Urban Outfitters use this technique to great effect to create relaxed and easy-to-shop retail environments that are in sync with how their consumer likes to shop—by spending time in-store and browsing new brands and products.

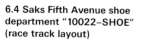

6.4 Saks Fifth Avenue shoe department "10022–SHOE" (race track layout)
A luxurious, comfortable, and spacious environment has been well planned to encourage customers to spend more time in-store; the customer's journey is interspersed with beautiful and welcome seating areas.

6.7

6.7 Urban Outfitters shoe area (free-form layout)
Shop floor space is maximized in this small format UK store; full-size runs of best-sellers are stored underneath a lead-in table. The advantages are that customers can "self-serve" and so sell-through is maximized as these styles are typically repeat purchases.

6.8 Converse flagship store in Soho, New York
High-impact display wall pulls customers into the Converse flagship store, New York City.

6.8

PLEND

This industry acronym describes the customer's ability to navigate his or her way around a store based upon the visual and physical cues that are designed to control the journey:

- **Paths**—walkways in between product zones (in department stores referred to as "mats") that should always be wide enough for wheelchair and stroller access.

- **Landmarks**—central display points that draw the customer into a central area. This could be a fixture with a higher-than-eye-level display that a customer is drawn to, and then can use as a reference point.

- **Edges**—these are perimeters, usually with wall fixtures that are permanent; however, some fixture systems are designed and sold to be fully flexible, so that configurations can moved and adjusted according to sell-through.

- **Nodes**—these are the intersection points; if these are best-seller zones, they are referred to as "hot spots" because they see the most traffic and are the most visible. These hot spots can also be used for clearing ends of lines as well as special purchases or promotions.

- **Districts**—also known as zones or mats, these areas house same/similar product types and are designed within the floor plan to flow logically from one product type to another.

6.9

6.9 PLEND
PLEND in action at Converse flagship store, Soho, New York.

Evaluating and Influencing Sales Performance

KPIs (key performance indicators) are measurable values that demonstrate how effectively a company is achieving its key business objectives. Marigay McKee (ex–chief merchant at Harrods and president of Saks Fifth Avenue) is succinct in her summary of the most important KPIs from a retail management perspective:

"It surprises me that most fashion buyers know everything there is to know about which trend and hemline we'll be wearing in six months but can't tell me what the density of their sales floor is, what the return is, what the dollars per square foot is, what their top-selling stock-keeping unit is, and how many times they've reordered it in the season. However fashionable the brand, we always start and finish with the numbers—the sell-throughs, the margins, the returns, the contributions—and then we talk about the pleasantries."

Additional KPIs are:
- Footfall—the number of people entering a store (or retail area) measured over a specific amount of time. Figures can be collated at regional, national, and international level and comparisons made, expressed as a percentage increase or decrease, e.g., -1.8% down year on year. Agencies such as Experian Footfall specialize in providing such retail intelligence services.
- ASP—stands for average selling price (at retail); the total revenue from a product's sales divided by the total number of units sold. The analysis of these figures and how they may change according to store location, region, time of year, etc. are vital in order to better understand customers, assist with product pricing, and sales planning and forecasting.
- ATV—stands for average transaction value and is the average customer purchase value, which again can be used to analyze store performance, planning, marketing, and promotion.
- Sales per square foot/meter
- Sales per linear foot/meter

Maximizing Sales Performance

Buyers, merchandisers, visual merchandisers, and retail personnel work together using the following techniques and devices to help maximize sales performance:
- Press pieces are displayed in windows with seasonal graphics or show cards depicting the press image and the publication. This same product should then feature ideally within the first 10% to 20% of the shop floor, again with show cards/graphics associated to magazine, etc.
- Impulse purchases should also be placed at the front of the store; these are often in seasonal highlight colors.
- Best-selling lines should be dual sited (or triple sited if enough space). These can be cross-merchandised with different product in order to look fresh and not repetitive.
- Nonseasonal core lines (often best sellers) can be placed in destination areas of the store, i.e., areas off the beaten path. These lines are referred to as "52 week" lines and are often a brand's most iconic style, e.g., the Hunter Original or the Ugg Australia Classic.
- Add-on sales are placed at POS (point of sale), e.g., shoe care products and socks.
- At POS, customer email details are captured and loyalty cards, etc. are promoted to encourage repeat purchases and customer loyalty.

RETAIL THEATER

In recent decades, the term "retail theatre" has emerged to describe the use of intended effects to improve customer satisfaction and loyalty (and therefore retail performance) by creating pleasant experiences for consumers (Baron et al. 2001).

Creating compelling, intriguing, and even traffic-stopping visual displays at retail is paramount to engaging the customer. There are many different components to this; store exteriors and windows, use of materials, color, composition lighting, music, and fragrance are all important elements. Retailers prioritize these differently depending on their own individual brand heritage and image. The AIDA (attention, interest, desire, action) model can be used to better understand this.

6.10 Liberty, London, UK
The Liberty of London Tudor building, completed in 1924, was constructed from the timbers of two ships: HMS Impregnable and HMS Hindustan.

The AIDA model

The AIDA model can be used to demonstrate how important customer engagement is via visually impactful retail displays. This applies to store design, exteriors, interiors, and use of props and graphics. For example:

- A—attention (awareness): grab the attention of the customer quickly so that the customer becomes more focused on the given window/display than anything else around him or her (the "traffic-stopping" effect).
- I—interest: raise customer interest by highlighting the benefits of a product. This can be done in an explicit way, or a more subtle way.
- D—desire: convince customers that they need this product, even if this is a luxury or nonessential item, that they can justify this purchase as it fulfills a need.
- A—action: prompt customers toward making a purchase or taking other action.

Store Exteriors and Windows

Retailers can adopt either a "glocal" or "global" approach (as described in Chapter 8), and this will be a major factor in determining what kind of approach is taken in different regions. There may be architectural considerations such as whether the building is protected, or a "listed" building, and this will in turn entail some constraints, such as use of materials, construction techniques, paint colors, etc.

A good example of AIDA in action is the Tuccia Prima atelier window where passersby were lured into the store by the spectacle created by placing a fully functioning mini-workshop in prime window space. Henri Bendel department store partnered with Italian trained master cobbler and founder of Tuccia Di Capri, Tove Nord, to create a new custom sandal brand, where handmade sandals could be designed, crafted, and collected within two hours from Henri Bendel's 5th Avenue flagship. Styles ranged from $190 to $445, and each week twenty pairs were created and received a Master Cobblers numbered stamp. The window atelier was highly effective and pulled customers into the iconic accessories store.

6.11

6.11 The Tuccia Prima atelier window at Henri Bendel, 5th Avenue flagship store
AIDA in action: attention, interest, desire, then action.

In Store

Continuing the theme of retail theater in-store, the smallest of details, seemingly not significant in isolation, take on a far greater significance when combined to convey a powerful message that appeals to the customer on a subconscious and conscious level, when designed and executed successfully. These include:

- Focal points. In-store displays with carefully arranged merchandise that immediately catches the eye. Works best in conjunction with a sight line (see Figure 6.12).
- Use of linear space, i.e., wall space. If key brands or strong products are positioned on linear fixtures, this will have a "magnet" effect and pull customers in, past other product areas, thus increasing time spent in-store.
- Lighting. Lighting plays a crucial role and can be a feature in itself, providing a central focal point of interest to lure customers in as well as enhance products.
- "Rule of three (or five)." A merchandising theory whereby asymmetric groupings of product are thought to be more aesthetically appealing as opposed to even numbers, typically twos in footwear, and this device can be effectively deployed at different market levels and different product types (see Figure 6.15).
- Signage and ticketing. A once overlooked area is gaining increasing focus, and witty and ironic executions add humor and character to the retail experience. Retailers are finding more creative and innovative ways of alerting the consumer to, for example, price promotions over and above traditional red ticketing and posters.
- Standards. Maintaining standards through consistency in terms of product quantities, positioning, and spacing in-store. Swedish retailer And Other Stories (part of the Hennes and Mauritz group) is particularly effective in this area, creating a very strong visual identity via simple, effective, and well-maintained stores that retain a boutique feel.

Key Considerations for Impactful Fashion Windows

- Themes and schemes
- Scale
- Props
- Color
- Lighting
- Signage and use of graphics

6.12

6.12 Sam Edelman collection area, Macy's
Use of colored props create a focal point at Sam Edelman's collection area, Macy's.

6.13 Eileen Fisher, Columbus Circle, New York City
Retailers keen to promote an ethical image favor the use of natural and reclaimed materials, such as Eileen Fisher in the United States. This aesthetic is subtly designed to appeal to an older demographic (40-plus).

6.13

6.14

6.14 Nicholas Kirkwood flagship store, New York City
The "rule of three" at Nicholas Kirkwood flagship store in New York; a good example of how appealing small, asymmetrical groups of product can look when merchandised together. This technique highlights choice in shapes and materials and "edits" the collection visually for the consumer to make the browsing and selection process easy and enjoyable.

CHALLENGES OF VISUAL MERCHANDISING FASHION FOOTWEAR

The visual presentation of footwear presents some quite specific challenges. The question of security when either pairs (e.g., Crocs and Havianas) or boxed product is placed on the shop floor to save on stockroom space (e.g., German discount footwear specialist You Know Shoes) means either higher staffing levels, to ensure that theft is kept to a minimum, or that specific "shrinkage" levels are planned for. Shrinkage, in retail terms, is where product is stolen, physically lost or lost due to paperwork errors, and is expressed as a percentage of turnover.

Using security tags on footwear can be appropriate in some instances, although this does detract from the aesthetic appeal of shoes and in most cases is not practical as product can be permanently damaged by various different tagging methods. Whereas clothing and accessories that are made from textiles can be tagged using devices with pins, this is not the case for leather where piercing will cause damage. Consequently, it is the industry norm for one shoe to be displayed on the shop floor.

Most retail centers/malls/locales will agree on only displaying either the left or right shoe as protocol, to prevent theft from two different stores (to make a pair). Organized thieves will travel distances in order to steal particularly expensive items if they know that protocols vary, particularly if the brand is a desirable one that they know they can sell on, or even steal-to-order.

6.15 Banana Republic, New York
A creative solution to displaying tagged ballet pumps in pairs at Banana Republic.

THE BLOCK HEEL PUMPS

Sound the triumphant horns - the block heel is back. Skip merrily through the upcoming months in the cool-to-be-kind shape of the season.

Nicholas Kirkwood Gold Block Heel Pumps, from Browns
Charlotte Olympia 'Eyes For You' Pumps, from Smets
Pollini Suede Pumps with Patterned Heel, from Browns

THE GO-GO BOOTS

Nancy Sinatra, eat your heart out. These Saint Laurent boots were made for swinging and twisting and popping and locking - and that's just what they'll do.

Saint Laurent 'Babies 40' Boots, from Eraldo
Saint Laurent 'Babies' Mid-calf Boots, from Al Duca d'Aosta

THE NEW MANNISH FLAT

What do you get when you cross masculine footwear with a costume jewellery designer? The dazzling creations of Martina Grasselli, whose sparkling brogues and loafers will make Coliac your new favourite shoe brand.

Coliac 'Adele' Gradient Moccasins, from Elite

THE RUBY SLIPPERS

Trust Dolce & Gabbana to dream up the season's most delectable fairy tale shoes. Just tap your heels three times and repeat after us: there's no place like home.

Dolce & Gabbana embellished heel Mary-Janes, from Parisi

THE WILD THINGS

Visual Impact at E-tail

New roles are emerging within the field of visual merchandising that are exclusive to digital platforms. These roles not only require advanced technical expertise but also require the digital visual merchandiser to collaborate with various departments to enhance the visual consistency of journeys to and within the company on all devices. Digital visual merchandising allows for changing product groupings far more quickly and efficiently than moves on the shop floor; different combinations of product can be tried and switched based on sell-through within hours and days, rather than weeks, so this allows for great flexibility when combining and presenting product. Visual themes can be presented by classifying and presenting product based on color, materials, silhouettes, designers, heel heights, size, and price. Assortments can be market tested quickly and efficiently, and space allocated to product increased or decreased based on sell-through on an hourly and daily basis.

As businesses evolve and combine purely "transactional" pages (where consumers can purchase product), editorial content, and social media, the role of digital visual merchandiser overlaps with merchandising and digital marketing roles.

Sophisticated examples of a combination of approaches are Farfetch.com (a digital platform for more than 400 premium boutiques globally) and Net-a-Porter.com.

6.16 Heel Surreal Shoe Guide at Farfetch
Farfetch employs creative styling techniques to enhance its digital visual merchandising.

CONSUMER EXPERIENCE–CUSTOMIZATION AND PERSONALIZATION

Product customization and personalization is an increasing trend favored to create consumer interest; whether part of a live window display and technical demonstrations, part of a seasonal promotion or ongoing service in-store, or part of an online offer, it is becoming ever popular and is evident at all market levels. Retailers and brands are able to use these projects as research mechanisms, just as much as commercial projects, to market test new products and price points with different demographics and psychographics. Their success can be measured by the increase in footfall, increase in sales turnover as well as average transaction value (ATV), and also consumer loyalty.

Converse and Havianas offer customization services in some of their flagship stores as part of an ongoing promotion coined MYOH (My Own Havianas) and a three-step process for Converse "Choose, Customize, Create," with single shoes to be embellished for $25 and $45 per pair.

WHOLESALE AND PR THEATER

In an increasingly competitive marketplace with brands trying to differentiate themselves from their competition, focus on visual presentation at wholesale (trade shows and showrooms) as well as at PR launches and events is becoming more important. These needn't be costly activities, with some highly effective wholesale presentations from upcoming and established brands evident at many trade shows.

The designer's inspiration and brand identity can be communicated through the careful selection and positioning of some colorful and charming props that reinforce the brand's image and creates a focal point that pulls buyers in to examine the range at closer quarters. Visual merchandising is also a crucial part of the public relations remit, as some visual merchandising creative agencies will produce solutions for retail, wholesale trade shows, and also for seasonal press launches.

6.18

6.17

6.17 MYOH (My Own Havianas)
Create and customize your own flip-flops MYOH at Havianas.

6.18 Keds stand at Bread and Butter Berlin trade show
Well-planned and effective use of color and props at Keds Bread and Butter Berlin, spring/summer 2015.

ETHICS IN ACTION:
The One Off Design Agency

Retail design companies are becoming increasingly aware of the need to design, build, and display in an ethical manner.

The One Off retail and branding design agency based in London, UK, works with the sustainable design team at Loughborough University (UK) to develop processes that allow for greener commercial design and the effective measurement of this process. It also partners with organizations such as the National Trust and the Design Council. The agency works across sectors, and its fashion clients include Crocs, Fat Face, and Speedo.

In working with Crocs, The One Off developed a new retail concept at Bluewater, Kent, UK. The 1,291-square-foot store builds on new merchandising techniques to promote new product where the traditional clog is not given primary space in-store, but instead new product categories take center stage. The goal was to try this concept in order for Crocs to align its visual merchandising across all store formats worldwide. The One Off also produced a handbook detailing its new visual merchandising strategies, setting the global standards for retail, and designed its European headquarters in Amsterdam one year later, in 2012.

6.19

6.20

6.19 Crocs store interior, Bluewater, Kent, UK
The Crocs store interior in Bluewater, Kent, designed by The One Off retail and branding agency.

6.20 Global visual merchandising guidelines
Global visual merchandising guidelines, by The One Off retail and branding agency, communicate visual merchandising standards and strategies to ensure consistency in all stores.

Industry Perspective:
Marc Debieux, Store Manager at Cheaney & Sons

Marc Debieux manages the Bow Lane store, in London's East End, and has worked in the footwear industry for twenty years, specializing in premium and luxury men's brands.

Can you describe a typical day in store?

Housekeeping is always the first duty of the day, followed by any stock work needed to be done. We will then contact any customers with outstanding orders before the lunch rush begins. We change our window displays every month, showcasing styles that reflect the season. We do themed windows using props and displays to attract attention, often collaborating with events in the local area.

How important is customer service and the retail experience when selling footwear?

Customer service is by far the most important aspect of retailing. The beauty of selling footwear is that every fitting is a consultation. You consult with the customer on their individual needs. Due to such a large amount of customer interaction, this creates a specialist retail experience only achieved in luxury footwear and bespoke tailoring.

How do you compete with shopping online? What are the challenges and how do you overcome them?

Online retailing is here to stay, and I think retailers need to embrace it. The beauty of luxury footwear is that fitting is essential to making a successful purchase. Different last shapes, width fittings, styles will all have an impact on the fit of a shoe. Also, working with leather will always create differences between pairs, be it in fit or the hand polishing that makes every pair unique. Because of this, it is still always favorable to try on in-store.

What online has taught us is that we need to make every visit an experience. I think it is so important that if a customer spends money, they should enjoy the purchase. We also offer regular shop events, "Scotch and Polish," which are free whisky tasting and shoes polished at our polishing bar.

What training and experience do you have?

I started my career with Clarks. I quickly became the men's floor supervisor before I was approached by Russell and Bromley. The training at Russell and Bromley was exceptional, with regular staff training sessions as well as outside programs. After three years I moved to Oliver Sweeney as assistant manager to open their new flagship store on the Kings Road. After three months I was promoted to the store manager in Middlesex Street (in the city). I was given the opportunity to work with the wholesale side of the business. This role included company trainer, where I wrote an extensive manual for all retail staff. I would conduct full training sessions both for individuals and groups, including Kurt Geiger, Selfridges, and Harrods. I also assisted with international development, traveling to the U.S.A. to help with trunk shows training at Saks and Nordstrom.

What is the best piece of career advice you have been given?

Become a concierge for your area; this will always bring appreciation and give you authority when selling. Always have a notepad to hand, take down details, and develop relationships.

SUMMARY

The businesses that have been able to identify and play to their strengths have done so by remaining true to their core values, and by designing and creating exciting and functional retail and e-tail spaces. This focus on excellent customer service in an easily navigable and comfortable retail environment has been crucial to their survival.

There are many practical reasons why displaying and retailing footwear differs from other product categories. Whether it be inherent security issues, stockroom space, additional staff needed to provide fitting and sales advice, or a lack of knowledge or expertise within this area, retailers are overcoming these challenges in creative and innovative ways that not only enhance the visual appeal of their selling spaces but also enhance sales turnover and profitability.

Although there are some practical considerations that set footwear apart, key visual merchandising principles can be employed to create exciting and eye-catching displays. These principles can be used in specialty stores and multiproduct stores alike, where footwear can be cross-merchandised to great effect to help a retailer or a brand tell a seasonal story. These techniques are the role of visual merchandisers; however, a visual merchandiser's role is only fully effective when considered as part of a process that includes retail designers, retail marketing managers, buyers, and merchandisers. These roles all support each other to design, create, manage, and maintain profitable sales space.

As some brands look toward designing and manufacturing product in a more ethical manner, it makes sense for the retail environment to also reflect these values. Cost-effective, recycled and up-cycled materials can be used to create highly effective sales spaces that are both visually appealing and flexible, enabling greater profitable while reflecting brand values.

As e-tail evolves, visual presentation online has moved away from the purely transactional nature of pages that merely sell product. In doing so, it demands a broader skills base from its management and personnel, in order to create a visual point of difference and identity on screen. Editorial content and additional services such as styling and personal shopping are also offered by e-tailers to remain competitive. Some successful fashion e-tailers are extending their brands beyond pure play (e-tail only) by including publishing and events as part of their omni-channel strategy. This is one of the fastest growing areas of the industry today, with job opportunities in these newly created roles accelerating rapidly.

DISCUSSION QUESTIONS

1. What are the most important factors in effective customer service in footwear today?
2. In your opinion, what are the five most important elements of effective footwear display at retail?
3. How does visual merchandising differ from high-end stores to fast fashion retail?
4. How can retail designers create more ecologically sound retail environments?
5. Why do e-tailers offer editorial content as well as transactional pages?

EXERCISES

• Select four leading fashion footwear specialty stores in your region (at different market levels) and discuss how their retail interiors have been designed in order to accommodate fitting services as well as product display. Discuss how effective this is and whether they can improve upon their retail design, and if so, how.

• By conducting primary research in a local department store that has the largest shoe area, draw an illustrated map of the area that contains the following information: zones by gender, i.e., men's, women's, and kids; which brands are placed where and how many SKUs are on display per brand; points of sale; fitting areas; locations of stockrooms; and identification of PLEND within the department.

• By analyzing the visual map you have constructed in the previous exercise, now identify and discuss the following: Which are the best-selling brands? Why? Where are the retail hot-spots? Where is the least desirable space on the shop floor and why? Is this retailer planning and using this space to its maximum advantage? What recommendations for improvement would you make?

• By conducting a content analysis of a successful mid-market and high-end e-tailer, what conclusions can you come to with regard to editorial space versus transactional space? Why have these websites been designed this way?

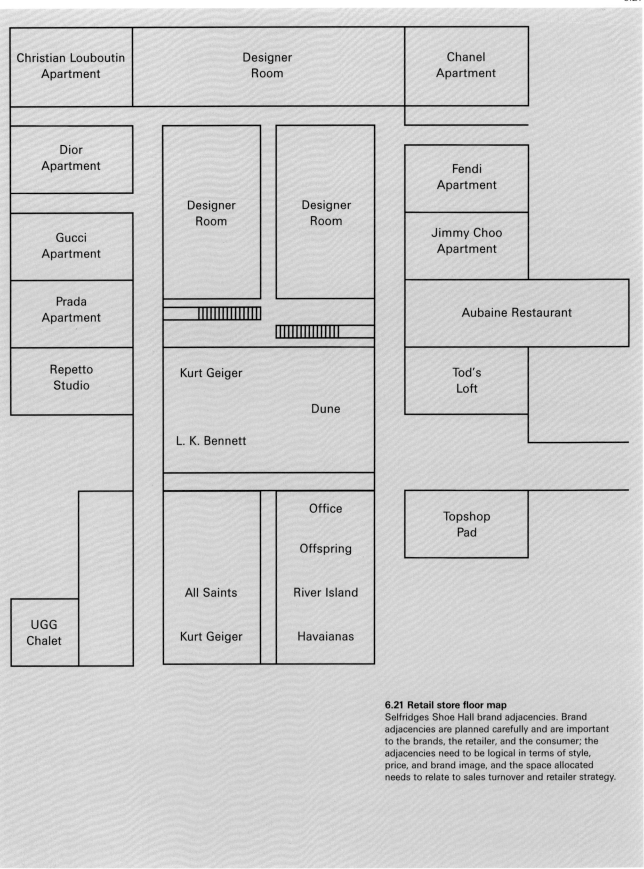

6.21 Retail store floor map
Selfridges Shoe Hall brand adjacencies. Brand adjacencies are planned carefully and are important to the brands, the retailer, and the consumer; the adjacencies need to be logical in terms of style, price, and brand image, and the space allocated needs to relate to sales turnover and retailer strategy.

KEY TERMS

Add-on sales—the sale of additional products or services to a customer at the time of purchase

AIDA model—a marketing acronym used to describe attention (or attract)/interest/desire/action

ASP—average selling price at retail

ATV—average transaction value at retail

Brand adjacencies—the placement of brands on the retail floor, or online, next to related brands

Core lines—best-selling product lines within a retail organization that represent a high proportion of sales turnover, and typically profit

Feature wall—a wall that utilizes product and marketing materials to create impact visually, to pull customers in

FMCG—fast-moving consumer goods

Footfall—the number of people entering a shop or shopping area measured in a given time frame

Free form layout—store floor plan most common in small specialty stores and within large stores' departments

Grid layout—store floor plan common in stores in which customers are expected to explore the whole store, such as grocery stores and drugstores

JTS—journey to sale

KPI—key performance indicator

Linear Space—wall units and other shelf space within a retail store, measured by feet or meters

PLEND—acronym for paths, landmarks, edges, nodes, and districts on the retail floor

Pop-up store—a retail store that is opened temporarily to take advantage of a trend or a seasonal product; can occur at different market levels

POS (point of sale)—location in store where a transaction is completed; the point at which a customer makes a payment in exchange for goods or services

Press pieces—key seasonal items selected by the brand and promoted in-store and via trade and consumer press

Race track layout—store floor plan common in large, upscale stores such as department stores where the customer is encouraged to pass through most areas by the strategic placement of product, fixtures, escalators, lifts, and exits

Show cards—in-store graphics merchandised next to, and promoting, seasonal press pieces

Shrinkage—the loss of inventory due to theft (external and internal), administrative error, vendor fraud, damage in transit or in store, and cashier errors; the difference between recorded and actual inventory

SKU—stock-keeping unit

07

BRAND IDENTITY AND PROTECTION

Learning Objectives

• Identify the key elements and theories of brand identity and equity for fashion footwear companies.

• Define the various categories of fashion footwear brands.

• Discuss how brands structure their brand portfolios via different trademarks to maximize their market share.

• Explain the strategies used by brands to protect their intellectual property and combat counterfeiting.

7.1 Luxury flagship stores in Hong Kong
Globally recognized and revered brand logos attract consumers and protect the brand.

INTRODUCTION

Fashion footwear is a complex industry where many variables converge to present brands with challenges that test their ability to become, and remain, credible. It is this credibility that is crucial for brands to survive, not just short term, but long term. And it is this credibility, in the eyes of the consumer, that allows brands to charge far in excess of production costs and overheads, and therefore yield greater profit.

Building brand identity and equity, consideration of intellectual property and legal aspects, as well as the brand–consumer relationship, are key to brand management. Sales and distribution strategies are also a critical part of this process (these are discussed further in Chapters 3 and 4).

Controlling brand image and identity is a process that is carefully planned and orchestrated across many different functions, or departments, within the company. It is not the role of just one function within the brand, but a shared responsibility that is often a fundamental part of a brand's strategy.

Brand protection is essential in order for brands to continue to maintain and grow their market share in an increasingly competitive marketplace. Consumers are more demanding and knowledgeable than ever before, so the choice of footwear goes way beyond the practical everyday needs and expresses how the consumer wishes to be perceived.

WHAT IS A BRAND?

First, we need to fully understand what a brand is, in order to be able to begin to understand how to manage and protect it. One of the first examples of branding dates as far back as 2700 BCE, as portrayed in ancient Egyptian hieroglyphics, where cattle owners would permanently mark their property via a hot branding iron and burn their sign or symbol into their livestock's hide. More famously, cowboys and cattle drivers in the Old West used these marks to identify the owner, protect cattle from rustlers (cattle thieves), and to separate them and track them in order to sell them.

A brand or sign is how retailers, and crucially, consumers, differentiate the same or similar looking products today. This early example reiterates how important branding is in today's busy marketplace. However, how a brand defines itself in today's crowded and complex commercial environment goes way beyond merely identification.

The Brand Wheel

Ogilvy defines a brand as being: "The intangible sum of a product's attributes: its name, packaging, and price, its history, its reputation, and the way it's advertised" (Ogilvy 1983). So by this definition we can begin to understand that many different departments within a branded business will be involved in the process; product design, marketing, supply chain, sales, public relations, and retail are just some of the functions that will all play a role.

7.2

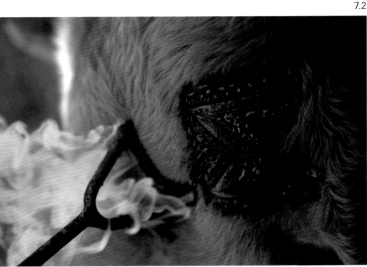

7.2 Cattle branding
Branding permanently marks the hide and dates as far back as 2700 BCE. Branding the hide of livestock denotes ownership, enabling identification and selling.

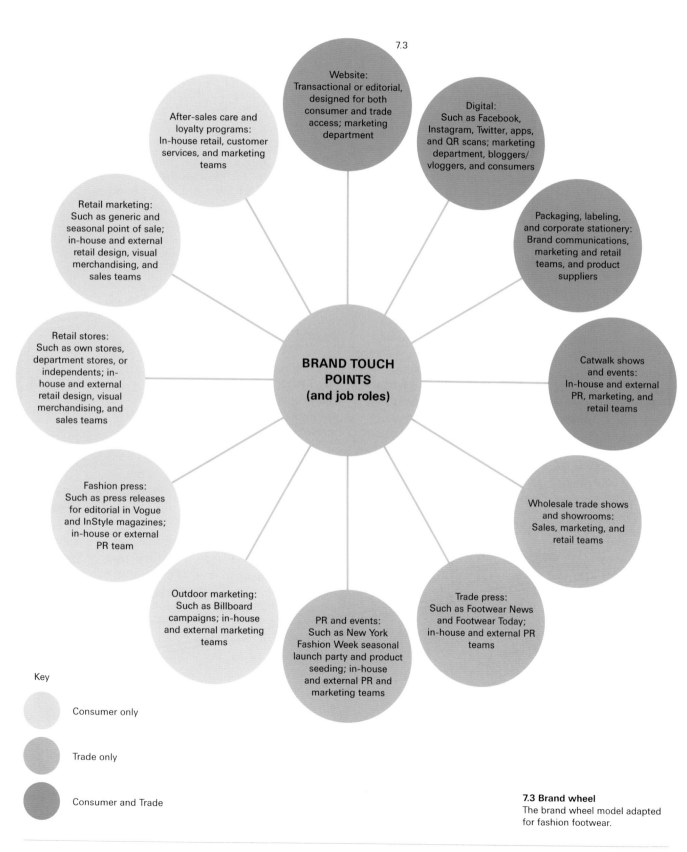

7.3

Website:
Transactional or editorial,
designed for both
consumer and trade
access; marketing
department

After-sales care and
loyalty programs:
In-house retail, customer
services, and marketing
teams

Digital:
Such as Facebook,
Instagram, Twitter, apps,
and QR scans; marketing
department, bloggers/
vloggers, and consumers

Retail marketing:
Such as generic and
seasonal point of sale;
in-house and external
retail design, visual
merchandising, and
sales teams

Packaging, labeling,
and corporate stationery:
Brand communications,
marketing and retail
teams, and product
suppliers

Retail stores:
Such as own stores,
department stores, or
independents; in-
house and external
retail design, visual
merchandising, and
sales teams

**BRAND TOUCH
POINTS
(and job roles)**

Catwalk shows
and events:
In-house and external
PR, marketing, and
retail teams

Fashion press:
Such as press releases
for editorial in Vogue
and InStyle magazines;
in-house or external
PR team

Wholesale trade shows
and showrooms:
Sales, marketing, and
retail teams

Outdoor marketing:
Such as Billboard
campaigns; in-house
and external marketing
teams

PR and events:
Such as New York
Fashion Week seasonal
launch party and product
seeding; in-house
and external PR and
marketing teams

Trade press:
Such as Footwear News
and Footwear Today;
in-house and external PR
teams

Key

Consumer only

Trade only

Consumer and Trade

7.3 Brand wheel
The brand wheel model adapted
for fashion footwear.

Brand Equity

Consumers are willing to pay more for branded fashion footwear because the brand will have built "equity" in their eyes. Brand equity is when consumers respond favorably to a brand by selecting it over its competitors and are willing to pay more for this product than unbranded goods. As such, brand equity can be regarded as an indicator of the success of a brand.

Keller (2013) defines brand equity as: "A brand has positive customer-based brand equity when consumers react more favorably to a product and the way it is marketed when the brand is identified than when it is not." This response depends upon a combination of recognition, associations, and judgments made by the consumer. Hence consumers are prepared to pay a much higher retail price for branded over unbranded. This is especially relevant with "designer" footwear where the cost of goods (cost to produce) is a very small percentage of the final retail price. At the high end of the market, this could be as little as 5 percent to 10 percent of the final retail price. This is because the consumer perceives the goods to be of a far higher value than their actual value due to what the brand represents. In the case of branded footwear, equity is earned over time, and this is built upon physical attributes such as quality and aesthetics, but also on more intangible qualities such as how the consumer wishes to be perceived and his or her values and beliefs.

Brand Personality

A brand's personality encompasses product (design and quality) and its communications (advertising and marketing). From a consumer perspective, the brand's personality should be easy to understand and instantly recognizable, even if the consumer cannot see the logo.

From a strategic perspective, a brand's personality can be augmented from within. This can be as part of a conscious plan to either maintain consumer perceptions of a brand or change these perceptions, often as part of a longer-term strategy. This is especially relevant if a brand's sales have decreased or they have lost market share, and the brand is actively planning how it can address this. We can better understand this by looking at the brand "onion" model.

BRAND ONION MODEL

The two outer layers are visible to the consumer, while the two inner layers are internal to the brand; for instance, the brand's strategic vision, as described by "a snapshot of the future," and the brand values, described as "a way of working and communicating."

If a brand is struggling and losing market share to its competition, or perhaps to other product categories, it will often invest in market research to identify who their current consumer is, and crucially who they could be. This market research is either conducted internally or externally, often by an agency, which will be briefed and given clear criteria to research. A new strategic direction can sometimes be the output of this kind of research (and how these goals are set and measured is explored in Chapter 9). Typically, the output of this kind of research, or "brand audit" could be to inform a new product design and marketing direction to appeal to new consumer groups in the future (see Chapter 9).

7.4

Vision
A snapshot of the future; a longer-term strategy, often supported by a strategic plan; can be measured via SMART goals.

Values
A belief system and way of working and communicating, often based on brand history and heritage and augmented from within; can be assessed internally, or by a third party (agency) when re-branding.

Personality
Informs all brand communications and touch points and influences consumer perception; can be measured by consumer research (qualitative and quantitative) and can be changed to inform a new marketing direction.

Positioning
Consumer perception in relation to the brand's competition, in the mind of the consumer; can be measured by consumer research and expressed as a perceptual map.

Key

Internal
These form part of the planned identity.

External
These are how the brand is seen by consumers.

7.4 The brand onion model
By applying the brand onion model to specific brands, we can better understand internal and external facets.

7.5 "Bally 1851" bespoke men's footwear at global flagship store, New Bond St., London
This made-to-order service was launched in 2014 at the opening of the New Bond St. store; an anatomically correct last is created and a selection of styles available with a made-to-order and exclusive made-to-color service, where customers have the option of customizing shoes to any color and intensity.

7.6 Manolo Blahnik collection area at Quartier 206, Berlin
Eponymous luxury brands' choice of retail partners is crucial; Manolo Blahnik partners with exclusive lifestyle department store Quartier 206 in Berlin.

CATEGORIES OF FOOTWEAR BRANDS

Different types of brands exist at different market levels, which can make definitions of types of brands complex; however, they can be categorized as follows:

- Bespoke brands: Shoes are handmade and fitted using the finest materials. A last is made to the customer's specific foot measurements and style selected. This kind of footwear may often be unbranded; however, established traditional and designer brands offer this service at the higher end of the market level, e.g., Bally from its made-to-order concept in London and Salvatore Ferragamo (made in Italy).

- Designer brands: This can be an overused term in the fashion industry. True designer brands operate at the highest level in the market and are often named after the designers themselves. The brand's specialism is footwear. For example, Nickolas Kirkwood and Manolo Blahnik.

- Secondary lines or "diffusion" brands: These are more affordable sub-brands aimed at a different consumer than the main, or "signature," line. They retain the brand's personality, and while they are often merchandised alongside the mainline in store, they are often sold to different retailers at different levels of distribution. This is another mechanism by which a brand can manage its image and reputation, by controlling its distribution. For example, Marc by Marc Jacobs and DKNY (by Donna Karan).

- Bridge brands: These are more accessible than designer brands. Their function is to bridge the gap between upper mid-market and designer level. They often have an extremely strong brand identity, as the consumer is being asked to pay more than for mid-market or private-label brands and needs a reason to do so. Strong marketing alluding to an aspirational lifestyle often plays a key role. For example, Tory Burch and Kat Maconie.

- Manufacturer brands: With an expertise in technology and innovation, these brands design, manufacture, and market their own products. Sometimes these brands will partner with chemical manufacturer brands and work in collaboration to produce technical fashion lifestyle product, such as Timberland (and Vibram).

- Private-label brands: These brands are owned by a retailer and are also known as store brands, retailer brands, or own label. These brands can operate across different retail channels such as department stores and multiples. Nordstrom's Halogen brand is sold exclusively in its stores, whereas Kurt Geiger's KG brand is sold in its own stores and department stores such as Selfridges and House of Fraser.

7.8

- Licensed brands: Licensing is an agreement whereby a brand owner sells the right to another company to use its name on branded merchandise. The company designs, develops, and markets the brand, for an agreed percentage. The company or "licensee" will have knowledge and expertise in the specific product sector; whereas the owner brand may not. For example, Pentland brands (based in the UK) are the worldwide licensees for Ted Baker footwear and Lacoste Chaussures.

- Brand collaborations: There are notably three main kinds of collaborations within fashion footwear. The first is where a footwear brand collaborates with a fashion, textile, or product designer (whose specialism is not footwear), for example Clarks (UK) working with the Irish textile designer Orla Kiely and print designers Eley Kishimoto, and Vans with Liberty of London. The second is where two footwear brands join forces, e.g., Jimmy Choo and Ugg Australia; the goal of this kind of partnership is to alter existing consumer perception and also attract a new consumer. And last, but not least, are retailer–celebrity collaborations where the celebrity is considered a "brand" with a specific personality and identity that is in sync with the retailer's, e.g., Sarah Jessica Parker for Nordstrom.

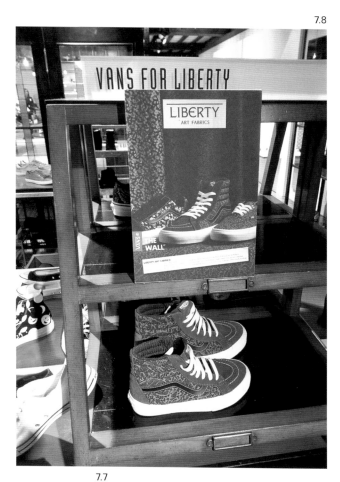

7.7

7.7 Tory Burch at flagship store in New York City
Use of the double-T logo creates a distinctive point of difference and is a feature of many best-selling styles.

7.8 Vans X Liberty London collaboration
Both parties gain from exposure to new consumers while still maintaining brand equity.

SEMIOTICS, BRAND ARCHITECTURE, AND TIERED BRANDING

Brands use different trademarks to maximize their market share by appealing to different consumer groups simultaneously. What these trademarks mean and represent to the consumer can operate at subconscious level. The brand's unique "visual handwriting" also plays an important role, i.e., all visual elements that support the logo such as specific colors and fonts. In order for a visual handwriting to be established and maintained by a brand, semiotics plays a key role here. Semiotics can be described as the theory and study of signs and symbols and their use and interpretation:

"'Commonsense' suggests that 'I' am a unique individual with a stable, unified identity and ideas of my own. Semiotics can help us to realize that such notions are created and maintained by our engagement with sign systems: our sense of identity is established through signs. We derive a sense of 'self' from drawing upon conventional, pre-existing repertoires of signs and codes . . . We are thus the subjects of our sign systems rather than being 'users' who are fully in control of them." (Chandler, 2006)

And so as consumers, we are not in control of the associations and connotations that we may glean from a company's logo design or use of color or imagery; instead, we are calling upon an already established visual "back catalogue" that helps us construct meaning. Adept marketers understand this process and use it to great effect as a device, as part of a strategy.

Brands can employ this device in differentiating one sub-brand from another, and this enables them to manage their brand "portfolio" by selling different product to different types of consumers with differing needs.

The synergy between the brand's and consumer's personality are major drivers to the purchase stage; consumers need to feel that the brand is speaking to them in a language that they understand. This becomes one of the key purchase triggers when consumers buy into a brand for the first time. And this recognition and

7.9

7.10

7.9 adidas "3 stripe" technical performance product
This product was originally designed for sports end-use.

7.10 adidas Originals fashion product
This product was designed for fashion end-use.

understanding deepens in time, and in doing so drives repeat purchases. It is important that consumers feel that they can relate to a brand on a personal level and that the brand's values are in sync with their own. In the case of adidas Originals, this brand has a loyal and devoted following spanning generations, with connotations of freedom, creativity, as well as originality that has developed over time via associations and partnerships with musicians and artists. The adidas Originals brand is distinctly different from the "3 stripe" brand, which focuses on sport.

The Brand Identity Prism

Jean-Noel Kapferer's brand identity prism (see Figure 7.11), first developed in 1992, can be used to explain some of the complexities of the brand personality and how the consumer relates to the brand, as a result of a constructed and strategically planned identity. The internal and external facets of the brand (see Figure 7.4) are depicted in both the brand onion model and Kapferer's brand identity prism. However, the brand–consumer dynamic is explored in more detail in Kapferer's prism.

In the brand identity prism, Kapferer identifies the following aspects of brand identity:

Brand physique: This is the brand essence expressed via physical features such as trademarks and iconic product that has come to represent the brand. For example, the Gucci double G logo and the horse bit loafer.

Brand personality: This relates to what personality the brand would assume if it were a person. How would it be described? What are the brand's attitudes or personality traits? For example, it could be said that the Sophia Webster brand is bold, extroverted, fun, feminine, and quirky (see Chapter 8 for Sophia Webster and J. Crew collaboration).

Brand culture: This is often intrinsically linked to a brand's history and heritage, and although it is often expressed externally, from an internal perspective it can mean how a company does business and what values the brand deems important. For example, Dr. Martens values integrity and individuality.

Customer self-image: This refers to the image that consumers have of themselves when purchasing and wearing the brand. This is often aspirational; it's who the consumers want to be and how the brand makes them feel. For example, a Jimmy Choo consumer wants to feel powerful, sexy, and elegant, and therefore selects this brand. Kapferer refers to this as the "internal mirror."

Customer reflected image: This is the image as portrayed in the brand's advertising, and it is the consumer's ideal version of him- or herself. For example, Jennifer Garner for Max Mara reflects the sophistication, classic elegance, and strength of the brand.

Relationship: This concerns social communication. What does the wearer want to communicate to the outside world? Through its adoption of this particular brand, what does the brand communicate and what does it communicate about the consumer? Does the consumer want to be seen as part of a style tribe? For example, the clog brand Swedish Hasbeens garners a loyal following on the basis that its wearers will be perceived as hip, artistic, and caring.

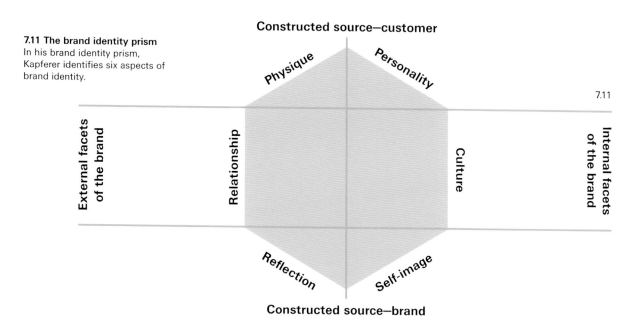

7.11 The brand identity prism
In his brand identity prism, Kapferer identifies six aspects of brand identity.

Constructed source–customer

Physique

Personality

7.11

External facets of the brand

Relationship

Culture

Internal facets of the brand

Reflection

Self-image

Constructed source–brand

7.12

7.12 Sophia Webster at Saks Fifth Avenue
The Sophia Webster brand is feminine, fun, and quirky and attracts a clientele who wish to be perceived as same.

Brand Architecture

Brand architecture can be defined as the way in which a brand, or "parent company," structures and organizes its main brands and sub-brands (its brand portfolio), and the way in which these brands relate to and are differentiated from each other. The goal of brand architecture is to maximize market share and shareholder value (if applicable). It can also be described as the combination of a brand's (or parent company's) history and heritage and its reaction to current marketplace challenges, in its home, as well as overseas, markets. In other words, how the brand adapts to external forces through the management of its sub-brands.

Brands need to be proactive in the management of their portfolios; in order to manage these, the synergies created between brands, the way in which these synergies leverage brand assets such as patents and sponsorships, and the role that each brand plays in the overall "brand mix" can all be considered part of brand architecture.

TIERED BRANDING STRATEGY

A brand (or parent company) can take a tiered approach to address different market levels and consumer groups simultaneously. This is particularly relevant in fashion footwear, as the parent company may have an expertise in particular materials and production techniques and will then adjust its pricing and marketing accordingly to meet with the consumer's needs. For example, Grendene (which specializes in plastic footwear, notably its famous "jelly" shoes) manages its Melissa and Impanema brands in this way, by operating at multiple market levels.

PROTECTING THE BRAND

Due to the significant revenues involved in branded footwear, it is necessary for brands to protect themselves. They can achieve this in different ways, and at different times; for example, when the brand is establishing its identity and as part of the ongoing management of their brand. In order to do this from a legal perspective, they have to establish and protect their intellectual property (IP). The fashion footwear industry is an IP-intensive industry, continually generating new designs and technologies that can be legally protected. IP is similar to any other type of physical property whereby under law it has a legal owner and therefore can be sold, bought, licensed, or damaged.

And while some aspects of IP can be viewed as intangible, such as know-how (which can include the talents, skill, and knowledge of the employees), training methods, technical processes, retailer lists, and distribution networks, these are very valuable assets.

What Is Intellectual property?

Intellectual property is highly relevant in brand management, as understanding what it is, why it is necessary, and how it can be used to protect a business is fundamental in creating a new brand identity and maintaining or changing an existing brand. The World Intellectual Property Organization (WIPO) defines IP as: "Intellectual property (IP) refers to creations of the mind, such as inventions; literary and artistic works; designs; and symbols, names and images used in commerce." It then goes on to explain different classifications of legal protection as follows: "IP is protected in law by, for example, patents, copyright and trademarks, which enable people to earn recognition or financial benefit from what they invent or create."

It is important to understand that while the WIPO is an international organization, it cannot enforce law in different countries, although it can recommend and canvass for a more global approach from a legal perspective. As such, a brand and its sub-brands need to ensure that they are fully protected in the correct "product class" (type of product) in every country or territory that it sells to and in. There are forty-five different product classes, according to the WIPO classification system

Intellectual Property—Some Basic Definitions

Patent: A patent is an exclusive right granted for an invention, which is a product or a process that provides a new way of doing something or offers a new technical solution to a problem. A patent provides protection for the invention to the owner of that patent for a limited period, generally twenty years.

Trademark: A trademark or brand name is a distinctive sign that identifies certain goods or services as those produced or provided by a specific person or enterprise. The period of protection for a trademark varies but can generally be renewed indefinitely.

Industrial design: An industrial design—or simply a design—is the ornamental or aesthetic aspect of an article produced by industry or handicraft; registration and renewals provide protection for, in most cases, up to fifteen years.

Copyright and related rights: Copyright is a legal term describing rights given to creators for their literary and artistic works (including computer software). Related rights are granted to performing artists, producers of sound recordings, and broadcasting organizations in their radio and television programs.

Source: http://www.wipo.int

(Nice Classification 2016); footwear and footwear-related items are within sections twenty-five and twenty-six. The Nice Classification (NCL), established by the Nice Agreement (1957), is an international classification of goods and services applied for the registration of marks.

IP Short- and Long-Term Factors

Designers and new start-ups, as well as established businesses, should be aware of the following:
- Ownership of rights (that have demonstrated a commercial return) is key in convincing future investors as to the commercial value of a company, or brand.
- Protecting IP enables designers to access new markets more safely through different modes of market entry, i.e., licensing, franchising, or entering joint ventures or other business models (bound by contracts). This includes overseas manufacturing, marketing, and distribution.

- IP rights are territorial, so a designer/brand should check that the trade name, trademark, and rights are available for use in all territories and product classes in which they intend to do business. This includes investigating IP issues before embarking on fashion shows or PR activity overseas, even before the sales process begins.

- Misuse (infringement) of the IP of others can be damaging and costly, via litigation.

Brands that seek to become global need to protect their IP in all the regions they sell in.

7.13

7.13 Coach
Luxury New York footwear and accessories brand Coach has stores in thirty-nine countries outside of the United States.

ETHICS IN ACTION:
Christian Louboutin versus Yves Saint Laurent

In the case of Christian Louboutin versus Yves Saint Laurent (2011) the U.S. district court expressed concerns that granting Louboutin a monopoly over the use of red outsoles would prevent fair competition. This case highlights the pitfalls of overreaching, and designers and brand owners need to carefully consider the scope of their rights before taking action.

**CHRISTIAN LOUBOUTIN S.A. et al., Plaintiffs,
v. YVES SAINT LAURENT AMERICA, INC. et al.,
Defendants.**
*United States District Court, S.D. New York
August 10, 2011*

Louboutin's claim to "the color red" is, without some limitation, overly broad and inconsistent with the scheme of trademark registration established by the Lanham Act. Presumably, if Louboutin were to succeed on its claim of trademark infringement, YSL and other designers would be prohibited from achieving those

stylistic goals. In this respect, Louboutin's ownership claim to a red outsole would hinder competition not only in high fashion shoes, but potentially in the markets for other women's wear articles as well. Designers of dresses, coats, bags, hats and gloves who may conceive a red shade for those articles with matching monochromatic shoes would face the shadow or reality of litigation in choosing bands of red to give expression to their ideas.

In sum, the Court cannot conceive that the Lanham Act could serve as the source of the broad spectrum of absurdities that would follow recognition of a trademark for the use of a single color for fashion items. Because the Court has serious doubts that Louboutin possesses a protectable mark, the Court finds that Louboutin cannot establish a likelihood that it will succeed on its claims for trademark infringement and unfair competition under the Lanham Act. Thus there is no warrant to grant injunctive relief on those claims.

GLOBAL COUNTERFEIT CULTURE

Counterfeit goods (also known as "fakes" or "knockoffs") are goods intentionally designed, produced, distributed, and marketed to mislead the consumer into believing that they are the genuine article. Counterfeiters will sell their goods at a significantly lower cost than the brand themselves, presenting a competitive threat to the brand. This threat is based on price rather than quality as often inferior materials and techniques are used, and so the markup (profit) is significant. The presence of these goods in the marketplace erodes the sales of the genuine article and damages brand equity and reputation.

The most popular premium and luxury brands are the brands that counterfeiters target, such as Prada, Gucci, Coach, Jimmy Choo, and YSL. Also, market leaders in sports and lifestyle categories such as adidas, Lacoste, and Vans are major targets for counterfeiters due to their market position and desirability.

7.14

U.S. Counterfeits

Due to its illicit nature, it is difficult to obtain the exact value of the global counterfeit footwear market. It can, however, be noted that that in 2013 the number of seizures in the United States of counterfeit footwear was 1,683, which was 214 less than in 2012. The estimated MSRP (manufacturer's suggested retail price) was valued at $54.9 million. The value of the seizures also dropped by almost 47 percent (from $103.4 million in 2012). Footwear had one of the largest declines over and above other product categories. So while the figure is still significant, years of focus in this particular product category are taking effect.

Why Must a Brand Protect Itself against Counterfeiters?

Counterfeiters produce inferior product using cheaper-quality raw materials and production techniques and will "de-spec" the original designs by simplifying them and omitting key design details. Sometimes consumers are aware that they are buying fake goods, although this is not always the case, and this confusion erodes brand equity. This is extremely damaging to brands both long term and short term as the consumers' trust has been damaged.

By offering IP protection through trademarks, copyright, design rights, and patents, innovation is encouraged and protected. If the brand is the owner of the IP rights, it is rewarded with exclusive rights and the platform from which to take action. If brands do not protect themselves by registering their IP in the territories they sell to and in, they are not legally protected; it is more difficult for them to take action against other parties, e.g., factories and suppliers producing the same or similar goods and selling them in these territories. This means lost revenue for the brand. Additionally, some consumers may prefer to buy the "fake" or counterfeit goods instead of the genuine goods, and this preference is always driven by price.

7.14 Christian Louboutin red outsole
Should a fashion brand be able to "own" a color?

CASE STUDY:
How Can Brands Protect Their Intellectual Property? UGG Australia

UGG Australia (owned by Decker's Outdoor Corporation) is the market leader in the sheepskin product category and has been one of the most imitated footwear brands in recent years. As such, it is one of the most proactive in taking action against those who threaten its business. One of the largest seizures of branded footwear in Europe was in 2011, at Southampton Docks. Thousands of pairs of fake UGG boots were confiscated, thought to be the largest ever seizure of its kind in Europe. More than 45,000 pairs of the boots were discovered by UK Border Agency officers inside six containers, with an estimated retail value of £9 million ($15 million). The containers originated from China and had been destined for Manchester, according to border force officials. The timing of this shipment was particularly strategic as it was shipped in early December, ready for the Christmas trading period.

UGG boots were originally worn by surfers to keep warm on the beach as the temperatures dropped, and to prevent injuries. The brand has evolved considerably since 1978, and affordable luxury and comfort are central to the UGG Australia brand. If these perceptions are damaged via the sale of counterfeit goods, the long-term effect would be to damage the brand's equity in the eyes of the consumer, making the brand less desirable.

High Street Imitations of Designer and Premium Brands

IP infringement does not just take place via counterfeiting. As with other product areas within the fashion industry, retailers and brands that operate at high street and mid-market level push the boundaries by imitating brands with higher brand equity. Sometimes this is done in a very subtle way, and in other cases there is little confusion as to who is being imitated. In the case of "Bobs," created by Sketchers, the resemblance to "Toms" was much debated in the blog and Twittersphere, and traces of the shoe were promptly removed from the Sketchers website. This is a constant challenge for brands such as Balenciaga, Jimmy Choo, Gucci, and Acne to protect their IP from retailers at lower market levels.

An important part of UGG Australia's strategy is consumer education—keeping one step ahead of counterfeiters by providing consumers who wish to buy the genuine article with practical tips on how to distinguish real from fake goods. Product and marketing materials are updated on a seasonal basis; not only are counterfeiters challenged to produce fake product, but fake marketing materials too. UGG Australia is keen to educate its customers using QR codes that can be tracked, as well as other elements such as labeling and staff training materials and programs for authorized dealers.

7.15

7.15 Fake UGG boots at retail
Counterfeit product based on best-selling styles retails at a fraction of the genuine article.

7.16

UGG
australia

ENSURING AUTHENTICITY.

We have added a new security label (behind sewn in label of the left boot or shoe), beginning with Fall 2013 product to help ensure that your new UGG® Australia product is authentic.

Follow the basic instructions below to confirm the authenticity of your purchase.

Locate your security label and turn it 90 degrees. The sun logos will change from black to white.

For more information about UGG® Australia, please scan the QR code found on your product label or visit uggaustralia.com.

7.17b

By taking this stance, the company is educating and informing its customers on how to differentiate between counterfeit product and the genuine article. It is also ensuring that by protecting its IP in the regions in which it sells, it can use this as a platform from which to take legal action, should its IP be infringed upon. While this activity in itself is labor intensive and costly, it allows UGG Australia to proactively manage its brand image and equity long term. By constantly evolving both its product and packaging (but still retaining the brand's personality), it can successfully control the brand image at many market levels.

WH15VMVS1C

7.16 Genuine UGG Australia product merchandised with branded shoe care kits at retail
Authorized dealers present the assortment with branded care kits and point-of-sale to express authenticity and drive sales.

7.17 QR scans and marketing elements reassure the customer of authenticity
A combination of technologies is employed to deter counterfeiters and reassure the customer.

7.17c

Industry Perspective:
Margaret Briffa, Founding Partner of Briffa

Margaret Briffa is an intellectual property and information technology lawyer and founding partner of Briffa.

What kind of training, qualifications, and experience would you advise someone to get if he or she wants to work in IP in footwear?

To be an IP lawyer you need to qualify as a lawyer first and then look to specialize after that. The legal qualification takes a minimum of five years. Jobs in intellectual property are very competitive and highly sought after. A real interest in the subject is a must and can be demonstrated through published papers and active involvement in intellectual property law groups. Also highly prized is an excellent grasp of the legal frameworks that touch on intellectual property law both in the UK and internationally, such as contract and commercial law. The work you do in a small niche practice with young vibrant businesses will be very different from the work in a large city practice working for multinationals and reporting to in-house lawyers. Find out what suits you early on and aim for it.

TM ® © **BRIFFA**

How does IP differ in footwear versus other product categories?

The law does not differ. It is the way the industry operates compared to other industries that does. Footwear, like many other industries, has a set sales cycle. Knowing the cycle helps you plan and get clients ready for action.

What advice would you give to a designer setting up his or her own footwear label with regard to protecting intellectual property?

They need to consider filing their distinctive brand name. They should search the trademark register to see what is already registered and also search on the internet for other brands that have similar names and could be a barrier to them gaining exclusivity in their chosen name. It is important to get this first filing right. A new business should take a look at the domain names it has and which it should capture to avoid others squatting on names as their reputation builds.

For upcoming designers and brands that are interested in international expansion, what are the first things they should do to protect their brand in a new territory?

As a general rule we advise that the portfolio extension is done gradually and that a business should only file as and when it is active or soon to be active in a particular country. The only exception to this is China where we recommend businesses file for trademark even if they have only just started. In China there are a lot of filings made by individuals looking for an opportunity to make money from brand owners by blocking the brand owners' own filing. The rules are such that once there is a filing in China, unless you can show that you have better rights because you have been selling there, it is impossible to get your brand back by following the legal process. This means that you either have to use a different brand for China when the time comes, or consider paying to recover your brand. Neither of those options is good. Both are hugely expensive compared to the cost of filing a trademark in China.

What is the single most valuable piece of career advice anyone has given you?

Put yourself in the clients "shoes." No, seriously. Imagine you are the business and the type of help and advice that would be useful to you, and deliver that.

SUMMARY

By exploring definitions of brand and branding and applying key marketing models, we can better understand how fashion footwear brands establish and protect their brand equity and differentiate themselves from their competition.

Brand protection is not a role exclusive to any one single job or department within a company; it is a shared responsibility. A credible brand personality cannot be invented overnight; instead it is born out of a brand's authentic history and heritage, and has a unique identity and personality of its own.

Through understanding the different types of footwear brands in the global marketplace and the differences in how these business models are structured, the role that they play within the global marketplace can be studied and assessed.

The importance of brand protection is paramount in brand management and enables brands to build sustainable long-term businesses while protecting themselves in their domestic and new international markets.

DISCUSSION QUESTIONS

1. What different departments and job roles are involved in brand management?
2. What are the advantages and disadvantages of private-label brands versus designer brands?
3. Who is the most influential fashion footwear brand in your region and why?
4. Using the same brand selected for question 3, describe the brand's personality, using the "brand onion" model.
5. Should a designer or a brand be able to own a color? Give reasons for your answer.
6. How can brands further protect themselves against counterfeiters? Are brands, governments, and NGOs taking a strong enough stance against this issue?

EXERCISES

- Select a market leader in a specific product category in your region and research how it manages its IP.

- Research and present back an example of a tiered branding strategy employed by one of the dominant footwear brands in your region. Is this strategy a global or glocal strategy? How successful is this? What recommendations would you make for the future?

- Taking the tiered branding strategy from the previous question, now apply Kapferer's brand identity prism model to both the main brand and a sub-brand to produce two applied examples. Compare and contrast the two.

KEY TERMS

Bespoke brand—shoes are handmade and fitted using the finest materials. A last is made to the customer's specific foot measurements

Brand architecture—how a brand, or "parent company," structures and organizes its main brands and sub-brands (its brand portfolio), and the way in which these brands relate to, and are differentiated from, each other

Brand collaboration—where a footwear brand collaborates with a fashion, textile, or product designer

Brand equity—when consumers respond favorably to a brand by selecting it over its competitors and are willing to pay more for this product than un-branded goods

Brand onion model—a theoretical model in which two outer layers are visible to the consumer, while the two inner layers are internal to the brand; for instance, the brand's strategic vision and goals

Brand personality—encompasses product (design and quality), and also its communications (advertising and marketing); how the customer perceives the brand

Bridge brand—more accessible than designer brands, its function is to bridge the gap between upper mid-market and designer level

Copyright—describes rights given to creators for their literary and artistic works

Counterfeit goods—goods intentionally designed, produced, distributed, and marketed to mislead the consumer into believing that they are the genuine article

Designer brand—operates at the highest level in the market; often named after the designers themselves

Industrial design—the ornamental or aesthetic aspect of an article produced by industry or handicraft

IP (intellectual property)— refers to creations of the mind, such as inventions; literary and artistic works; designs; and symbols, names, and images used in commerce

Kapferer's brand identity prism—a theoretical model used to explain the complexities of the brand personality and how the consumer relates to the brand

Licensed brand—an agreement whereby a brand owner sells the right to another company to use its name on branded merchandise. The company designs, develops, and markets the brand, for an agreed percentage

Manufacturer brand—with an expertise in technology and innovation, these brands design, manufacture, and market their own products

Private-label brand—owned by a retailer; also known as store brands, retailer brands, or "own label"

Secondary lines or diffusion brand—more affordable sub-brands aimed at a different consumer than the main, or "signature," line

Semiotics—the theory and study of signs and symbols and their use and interpretation

Tiered branding—strategy to address different market levels and consumer groups simultaneously

Trademark—a distinctive sign that identifies certain goods or services as those produced or provided by a specific person or enterprise

08

BRAND MANAGEMENT

Learning Objectives

- Explain how brands evolve and are managed by discussing how consumer awareness, perception, and adoption are translated into sales and how successful brands appeal to style leaders to late adopters simultaneously.
- Discuss the dynamics of brand collaborations by identifying elements for success.
- Define what makes a footwear brand iconic, via the exploration of a brand's evolution, and how functional brands become fashion brands.
- Explore the motivations that shape social enterprise and establish an understanding of how these motivations shape brand vision.

8.1 Agyness Deyn for Dr. Martens wearing the Langston boot (photograph by Gavin Watson)
The AW11 "First and Forever" campaign expresses the brand's heritage and values.

INTRODUCTION

Successful fashion footwear brands are not born overnight but rather evolve over time as a result of unique product, in-depth consumer research, carefully planned strategies and in many cases by capturing the spirit of the time, often by chance. This chapter explores how brands morph from "fad" brands into truly iconic brands by appealing to many different consumer types simultaneously.

Collaborations have become an increasingly popular and effective way for brands to manage their image and reputation long term and to maintain a sense of identity for their existing consumers, while attracting new consumers. Factors for successful collaborations are also highly strategic; considerations are which brand to partner with and why, whether the partnership will be mutually beneficial, whether both parties' image and reputation will be enhanced, and how the collaboration will tie in with broader strategic aims.

Not all brands have the same motivations. Profit, prestige, social enterprise, and combinations thereof are just some of the factors that drive behaviors. From a marketing and brand management perspective, these primary motivating factors are critical in determining advertising campaigns, sales distribution, retail strategy, and design. These motivating factors can either remain the same throughout a brand's life span or shift over time as the brand changes and evolves. These changes can be brought about by a change of ownership, key personnel, or product direction due to market demands.

To be deemed an "iconic" brand is indicative of many factors; it is often associated with technical expertise and being the first to market, or market leader, within a particular product category. To make an iconic shoe or footwear brand, it is necessary to have timeless appeal, to transcend seasonality, to have functional roots or a rich and authentic heritage, and crucially to have powerful cultural connotations. Possessing, recognizing and communicating these attributes are an intrinsic part of successful brand management.

WHAT IS BRAND MANAGEMENT?

Brand management can be described as a total approach that encompasses the control and planning of goods, services, marketing, advertising, and distribution in order to maintain and improve results. In today's rapidly changing and evolving commercial environment, brands need to be both instantly recognizable and nimble enough to change should the market or local market demand this. If a brand is to adopt a "global" approach, i.e., that product, and every aspect of the promotional mix, is the same world over; consistency is key. A global approach is considered highly effective when a brand is establishing itself and building its identity.

As explained by Easey (2009), "If every aspect of the promotional mix is integrated then the effect is likely to be stronger and long lasting, as the imagery and treatment i.e. the hand writing is consistent every time it is used— be it in an advertisement, on the web, in a store, on a letterhead, it becomes distinctive and is: Recognisable; Repeated; Reinforced; Reiterated; Recalled."

When a brand is well established, it may decide to vary its approach for different regions. A global approach may not be suitable or relevant in all instances, and a "glocal" approach is favored instead. Glocal is a hybrid of global and local and takes into consideration both the overarching brand message and identity and the specific demands of a particular region. For example, this could be regionally specific product, marketing, and even distribution. The demands and constraints of the local market could be due to key events that take place within the region, dominant consumer groups, and competitive elements also. A branded footwear business is better placed to manage its distribution if it understands the idiosyncrasies of each country that it sells in (this is further explored in Chapter 6).

**8.2 Bally, global flagship store,
New Bond Street, London**
Interior designed by David Chipperfield architects and launch event organized by Modus PR and branding agency.

Glocal Marketing Approach

The Swiss premium footwear and leather goods brand Bally employs a glocal marketing approach highly effectively. By partnering with key institutions in strategically selected fashion cities globally, it is able to inform and educate existing and emerging markets about the brand while maintaining brand values such as an appreciation of art and craftsmanship.

In 2012 Bally partnered with Vogue to celebrate Vogue's "Fashion's Night Out," an initiative that coincides with the key fashion weeks in Berlin, Madrid, Paris, Milan, London, New York, Los Angeles, Costa Mesa, Sidney, Shanghai, Dusseldorf, Istanbul, Amsterdam, and Rome.

The "Brilliance" ballerina flat was created in an exclusive color to celebrate the event, and distribution was limited to Bally stores in these cities only, creating demand and brand heat at an extremely important time in the fashion year calendar. This is a particularly effective strategy as it speaks to consumers and the industry simultaneously.

Bally's new London store opened in 2014, where it tried out a new bespoke men's concept "Bally 1851,"offering customers the opportunity to purchase bespoke made-to-order to shoes up to a retail value of £20,000.

8.2

MEASURING SUCCESS—AWARENESS AND ADOPTION

To decipher how successful a footwear brand is, or could be, measuring the value of the brand is necessary. This involves the measurement of both tangible and intangible variables. Tangible factors including income, potential income, share price, staff retention, and recruitment are factors that can be checked and tracked. And intangible factors that can be more complex to measure, such as consumer recognition and loyalty, can also be researched and evaluated.

Brand equity (as discussed in Chapter 7) is an integral part of a brand's value. These two terms are often used as synonyms; however, for the purposes of understanding successful brand management, it is important to separate them. Brand value refers to the financial value of a brand; brand equity is just one facet of how a brand's value can be determined. It is also important to note that both concepts do not always operate in tandem, e. g., a sharp increase in sales revenue as a result of widening distribution may temporarily increase sales and shareholder confidence, but at the same time, brand equity can be eroded because the consumers' perception of the brand has been undermined, creating a longer-term problem for the brand. Successful brand management relies on maintaining the delicate balance between these two principles, a healthy balance sheet, and desirable product in the eyes of the consumer.

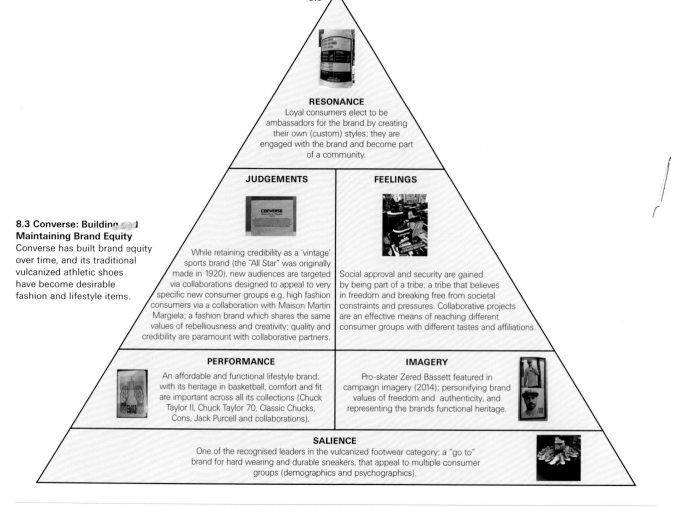

8.3

RESONANCE
Loyal consumers elect to be ambassadors for the brand by creating their own (custom) styles; they are engaged with the brand and become part of a community.

JUDGEMENTS

While retaining credibility as a 'vintage' sports brand (the "All Star" was originally made in 1920), new audiences are targeted via collaborations designed to appeal to very specific new consumer groups e.g. high fashion consumers via a collaboration with Maison Martin Margiela; a fashion brand which shares the same values of rebelliousness and creativity; quality and credibility are paramount with collaborative partners.

FEELINGS

Social approval and security are gained by being part of a tribe; a tribe that believes in freedom and breaking free from societal constraints and pressures. Collaborative projects are an effective means of reaching different consumer groups with different tastes and affiliations.

PERFORMANCE

An affordable and functional lifestyle brand; with its heritage in basketball, comfort and fit are important across all its collections (Chuck Taylor II, Chuck Taylor 70, Classic Chucks, Cons, Jack Purcell and collaborations).

IMAGERY

Pro-skater Zered Bassett featured in campaign imagery (2014); personifying brand values of freedom and authenticity, and representing the brands functional heritage.

SALIENCE
One of the recognised leaders in the vulcanized footwear category; a "go to" brand for hard wearing and durable sneakers, that appeal to multiple consumer groups (demographics and psychographics).

8.3 Converse: Building and Maintaining Brand Equity
Converse has built brand equity over time, and its traditional vulcanized athletic shoes have become desirable fashion and lifestyle items.

Customer-Based Brand Equity and Consumer Adoption

Keller (1993) defines customer-based brand equity as:

"Customer-based brand equity is defined as the differential effect of brand knowledge on consumer response to the marketing of the brand. A brand is said to have positive (negative) customer-based brand equity when consumers react more (less) favorably to an element of the marketing mix for the brand than they do to the same marketing mix element when it is attributed to a fictitiously named or unnamed version of the product or service . . . Customer-based brand equity occurs when the consumer is familiar with the brand and holds some favorable, strong and unique brand associations in memory."

Converse is a good example of how brand equity has been built over time and is maintained by careful and strategic brand management. This model demonstrates how brands build equity first through establishing recognition and relevance (salience), working toward the ultimate goal of resonance through loyalty and attachment, when consumers become advocates of the brand. Bearing in mind that Converse first named its canvas basketball shoe the "All Star" in 1920, this has been a long and gradual process of building brand equity in the eyes of the consumer, to create a brand with longevity. When considering how consumers first adopt new trends, Roger's diffusion of innovation theory, as discussed in Chapter 1, is also relevant. It is also pertinent to consider the time that it takes a fashion shoe (or brand) to go from being a desirable commodity to being categorized as a "one-season-wonder." If this process is too accelerated, a brand can "burn out." This is a result of the brand's equity not being established enough to withstand external and internal pressures, such as increasing distribution too quickly and an overreliance on one or a few key models.

One such example is the Crocs brand; having sold in excess of 300 million pairs globally since its inception in 2002, few brands polarize opinion quite like Crocs. Made from a patented foam resin, Croslite, these shoes became a fashion phenomenon in 2006 and 2007. Global sales peaked at $847 million in 2007. However, the business was overreliant on one style, the original clog, which accounted

8.4 Jared Leto wears Crocs
Product placement in 2006 is successful as Jared Leto is photographed wearing the brand at New York Fashion Week.

for almost a third of revenues in 2007. Not only was the brand overreliant on one style, but it was over-distributed, and—compounded by the start of the global financial crisis in 2008—Crocs' shares crashed from a peak of $68 to less than $1 as the firm went from a $168 million profit in 2007 to a $185 million loss the following year.

Struggling to survive, the brand restructured by changing its management, cutting its workforce by 2,000, and closing factories in Brazil and Canada as well as controlling its distribution more tightly via changing its distributors and creating different product ranges for different retail channels. By segmenting its market through analyzing what its consumers' behavioral, demographic, psychographic, and geographical differences were and designing product and targeting retailers aligned to these consumer needs, Crocs has been able to rebuild its business, post-2008.

WHAT MAKES SUCCESSFUL BRAND COLLABORATION?

Brand collaborations in fashion footwear can be defined as two or more parties working together in the design, production, and marketing of co-branded product. Their goal is to enhance existing perceptions of the brands by challenging the consumer to think of them in a different way by combining the different perceived properties within a single product, or range. These collaborative partnerships are often short term (most usually for one season), but with longer-term aims and objectives.

When the strength of two (or more) brands is combined, the premium that consumers are willing to pay is increased and the brand's intellectual property is also strengthened by becoming more resistant to copying by private-label manufacturers and counterfeiters.

Collaborations have become an increasingly popular way for brands to keep their existing consumers' interest and loyalty by reinforcing core brand values (as discussed in Chapter 7) while appealing to new consumers. This is a delicate process that involves a certain amount of risk; by their very nature, brand collaborations are highly visible, and therefore they need to be carefully managed to ensure that the image and reputation of all parties are enhanced and that there are both short-term and long-term strategic benefits.

Brand Collaboration

Brand partnerships are part of a broader strategy. From a commercial perspective, the result of consumer perception being altered can help to:

- Increase sales turnover by attracting a new consumer.

- Re-weight the annual order book to decrease the dependence on one season versus another (e.g., reduce the dependence on spring/summer versus autumn/winter.

- Re-balance the annual/seasonal order book by gender, i.e., increase one gender to create a more balanced order book.

- Reduce the percentage of a single style or product category, thus decreasing the risk should the style or category become obsolete or unfashionable.

- Reduce the percentage of a style or product category that uses a particular material that may be in short supply or subject to rapidly fluctuating price changes.

" SOPHIA'S INTERESTING . . . IN THE WAY SHE SEES PRINTS AND IS SO BOLD IN HER CHOICES, REALLY FEARLESS AND LIVES A DREAM STATE IN THE WAY SHE DESIGNS . . . AT THE SAME TIME WE AT J. CREW DON'T NECESSARILY HAVE THAT POINT OF VIEW SO THE COMBINATION OF WHAT WE'VE CREATED IS SOMETHING I THINK NEITHER ONE OF US WOULD COME UP WITH. WHICH IS TO ME THE POINT OF A COLLABORATION."
JENNA LYONS, J. CREW PRESIDENT AND EXECUTIVE CREATIVE DIRECTOR, ON SPRING/SUMMER 2014 COLLABORATION WITH DESIGNER SOPHIA WEBSTER

Different Types of Brand Collaborations

While brand synergies are often shared in successful collaborations, in order for consumer perceptions to be challenged positively, the project will often bring an element of the unexpected, unique, and exclusive, with distribution limited and tightly controlled. The rationales are to market test the idea via controlled and limited sales numbers and to manage risk. If successful, the projects create brand "heat" (i.e., very positive publicity) and if unsuccessful should not compromise the core existing business from a commercial perspective.

Such projects exist across many market levels and can be categorized as follows:

Premium shoe brand and luxury shoe brand: UGG Australia and Jimmy Choo worked together in autumn/winter 2010 on a limited-edition collection of women's sheepskin footwear. The collection of five styles, made in multiple colorways, was priced between $495 and $795 (and priced between £495 and £695 in the UK) and sold at UGG Australia's and Jimmy Choo's own retail stores and a selection of prestigious department stores, which were already key retail partners for both brands, e.g., Nordstrom. The price architecture was designed to sit in between the ceiling price point for UGG Australia and the entry and mid-price point for Jimmy Choo, creating an enhanced sense of luxury for UGG customers and accessibility for Jimmy Choo customers, as well as cementing already strong press and retailer relationships. The advertising campaign featured model Amber Valletta and was shot by Inez Van Lamsweerde and Vinoodh Matadin.

Premium retailer and celebrity: Nordstrom and Sarah Jessica Parker, together with Manolo Blahnik chief executive George Malkemus, launched a twenty-five-style collection in spring/summer 2014. All styles were made in Italy and priced between $195 and $485 (with an average price of $300). The line was available only at Nordstrom's key twenty-five footwear stores and created excellent publicity by reinforcing shared brand values.

Luxury shoe designer and mid-market fashion brand: British designer Sophia Webster and J. Crew partnered in late summer 2014 to produce a capsule collection of fifteen styles (of pumps and mid-heels) with price points from $320–$695. The range was available in selected stores and online and helped promote the British designer's reputation in the United States (enhancing her existing distribution in selected Saks Fifth Avenue, Bergdorf Goodman, Neiman Marcus, and Nordstrom stores). The price points were lower than Webster's main range, creating a sense of accessibility for new consumers, particularly in the United States while still reinforcing J. Crew's market position of affordable style (see Chapter 7 for Sophia Webster and brand personality).

Premium shoe designer and mid-market footwear chain: British designer Kat Maconie and British footwear retailer Dune created a capsule collection of three shoe styles and one clutch bag. The collection was launched in spring/summer 2012 and was inspired by Miami Art Deco style. Prices ranged from £75 to £135 ($128–$230). Kat Maconie, whose shoes fuse high fashion with design function, are favored by celebrities such as Sienna Miller and Rosie Huntington-Whiteley. Kat Maconie said of this collaboration "The collection has a slightly sexier and glamorous feel to it than my own collection, which tend to be more subtle and understated. The heels are higher and thinner—I normally like to balance my chunky heels with strong architectural uppers but have gone for a slightly lighter, sexier detail in some of the Dune designs." The inspirations, and the designs themselves, remained true to the British designer's aesthetic, while extending the brand's distribution by making it more accessible to a wider audience via high street stores at lower price points. Dune was also able to enhance its brand image by adding an upscale up-and-coming young designer to its portfolio.

8.5

Mid-market footwear brand and high-end fashion designers: The Brazilian brand Melissa (owned by Grendene) has made designer collaborations such an integral part of its ongoing strategy that it has worked with Vivienne Westwood (since 2008) and Karl Lagerfeld and Jason Wu on an ongoing basis. These ongoing licensing partnerships are also an important part of the designers' overall strategy by being an accessible and affordable face of cross-category empires spanning apparel, footwear, accessories, jewelry, eyewear, and fragrance. The main the footwear collaborations are with Grendene exclusively, although this is not always the case due to the power of the designer brands and their ability to negotiate directly with retailers, e.g., Karl Lagerfeld for Hennes.

Mid-market footwear brand and architects: Architect Zaha Hadid has partnered with several footwear brands in recent years. Designing for Lacoste Chasseurs (owned by Pentland brands, based in London) in 2008, a premium mini-range with a maximum production run of 1,000 pairs were sold at Colette in Paris, Dover Street Market in London and 10 Corso Como in Milan. This collection went on to inspire a diffusion range of 10,000 pairs, thus cascading the project down through different levels of distribution. The architect also collaborated with Melissa in 2008, designing women's product using advanced digital modeling techniques. The combination of Melissa's plastic injection mold technology lent itself effectively to the fluid lines of these designs producing highly innovative product, which also allowed the architect to experiment and create in another 3-D medium.

In 2013 the architect collaborated with Rem D. Koolhaas, creative director of United Nude, to create an innovative and technically advanced shoe called the "Nova," which was launched in July at L'eclaireur, Paris. Hadid developed a cantilevered system that allowed the 16 centimeter (6.25 inch) heel to appear completely unsupported. Techniques such as injection and rotation molding, as well as hand-molded methods such as vacuum casting, were combined. There were very few colorways, or finishes, and each finish was limited to 100 pairs. Retail prices ranged from £1,373–£1,868 ($2,340–$3,180). All three brands were able to enjoy the "halo" effect of being associated with the industry-leading and award-winning architect.

8.6

8.5–8.6 Original sketches for Dune collaboration
Kat Maconie collaborated with multiple retailer "Dune" to launch a collaboration in spring/summer 2012. Both parties gained exposure to new consumers.

Mass-market retailers and high-end fashion designers:
Target and Peter Pilotto worked together in spring 2014 on a clothing, accessories, and footwear range. The unique and distinctive digital prints that are design duo Peter Pilotto and Christopher de Vos's signature style were applied across all product categories to create some of the most accessible "designer" items. The mini-range featured vulcanised slip-on and lace-up styles in a numerous prints and color ways, and retailed for $29.99 in the United States at selected stores and online. They also retailed at £30 ($50) in the UK, available via Net-a-Porter, thus protecting the brand's premium image by providing alternative distribution that did not conflict or compete with Target.

Hennes has famously collaborated with a number of high-end fashion designers since its first project with Karl Largerfeld in 2004. Footwear collaborations have included Jimmy Choo (autumn/winter 2009), Maison Martin Margiela for autumn/winter 2012, Isabel Marant for autumn/winter 2013, and Alexander Wang for autumn/winter 2014. Distribution is carefully managed to top stores only, with ranges selling out within hours due to the power of the designer brands and the extensive PR and advertising campaigns that start six to eight months ahead of launch.

Mass-market footwear brands and fashion designers:
In spring/summer 2014, Aldo launched a collection with Susanne Ostwald and Ingvar Helgason, the German-Icelandic couple behind London-based label, Ostwald Helgason. Following up on four successful collaborations with Preen, the "Rise" initiative became the permanent banner for their high-fashion collaborative projects, the first of which was in 2010. The label is known for its signature use of color and bold stripes, as well as its modern take on luxury sportswear, and price points are pitched from $120–$130. The retailer was able to enjoy publicity at both New York and London Fashion Weeks, while the London-based design duo enjoyed more exposure in the U.S. market.
Mass-market fashion brands and niche footwear brands: Hennes and Swedish Hasbeens worked together in summer 2011 and launched their three-style collection in 150 stores worldwide. The collection was exclusive to Hennes with prices ranging from $68–$82 (€49,95–€59,95). The styles did not deviate from their distinctive 1970s aesthetic, and with the brand's cult following (being worn by Hollywood "A" listers such as Sarah Jessica Parker and Maggie Gyllenhaal), bohemian and ethical connotations were created as a result of this partnership.

8.7 Melissa and Vivienne Westwood and Karl Lagerfeld collaboration
Grendene-owned brand Melissa has ongoing partnerships with Vivienne Westwood and launched a collaboration with Karl Lagerfeld in 2013, which was a four-season contract.

Mass-market footwear brands and textile designers: British comfort brand Clarks worked with British/Japanese label Eley Kishimoto in spring/summer 2013 featuring its unique prints. A range of pumps, wedges, and desert boots were all produced in three distinctive prints and were available at larger stores and online only. Due to the iconic nature of the silhouettes (the original Clarks desert boot in particular) and the prints, sell-through was rapid and the positive PR generated enhanced Clark's image as being a fashion-forward, as well as good-quality, comfort brand and brought the work of Eley Kishimoto to a wider audience.

In spring/summer 2014, Clarks collaborated with Irish textile and clothing designer Orla Kiely. The six-style collection was Kiely's first footwear project, and Clarks enjoyed the benefit of launching this collaboration at London Fashion week, alongside Kiely's apparel and accessories line, while Kiely's brand was further enhanced by having a trusted and established partner to launch with, rather than taking the licensing route.

Mass-market retailers and celebrities: British fast fashion retailer New Look and actress and model Kelly Brook first collaborated in 2012 on apparel and accessories. The range was heavily influenced by 1940s and 1950s vintage styles with raffia wedges, peep toes, bows, and block color accents. Price points were £14.99–£39.99 ($25–$68). New Look has 1,160 stores worldwide in 120 countries, so this was beneficial in promoting Brook's global profile, while enhancing New Look's casual high-fashion image, by adding accessible glamour appealing to a wider demographic.

Harris Tweed: The traditional handwoven cloth of the Outer Hebrides of Scotland is deserving of its own category. This robust and attractive cloth has been used by many footwear brands in recent years for seasonal projects and ongoing partnerships. The following brands have all collaborated with this most authentic of materials: Clarks, Dr. Martens, Mandarina Shoes, Jaggy Nettle, F-Troupe, Vans, Nicholas Deakins, Converse, and Alfred Sargent and Charles Tyrwhitt. Protected by an act of parliament in 1993, Harris Tweed must be dyed, blended, carded, spun, warped, woven, finished, examined, and stamped only in the Scottish Outer Hebrides by local crofters and artisans. Partnership brands benefit from this strictly controlled process and by association are deemed to represent honesty, authenticity, and a timeless rugged appeal.

8.8

8.8 Clarks and Eley Kishimoto collaboration
Eye-catching and distinctive product and packaging; Eley Kishimoto signature print "Flash" in black and white is used for dust bags.

THE CREATION OF A FOOTWEAR ICON

The following attributes are an integral part of the style DNA of an iconic shoe or footwear brand: timeless appeal, ability to transcend seasonality, functional roots or a rich and authentic heritage, and crucially powerful cultural connotations. And if many different types of consumers, or "style tribes," adopt the style, this in turn adds to its status as an icon. Iconic styles may be fashionable, but they often represent "anti-fashion" and their appeal is what and whom they represent. (Please also see Hunter case study in Chapter 9.)

From Function to Fashion

Bronislaw Malinowski (1884–1942), the founder of functionalism, used the term needs functionalism. His form of functionalism focused on the individual satisfying the basic seven needs of humans, which include nutrition, reproduction, bodily comforts, safety, movement, health, and growth (Moore 2009). Functional footwear addresses most of these needs, and examples of iconic footwear brands that are born from humble functional origins are Scholl and UGG Australia, originally being worn for medical purposes and by surfers for warmth after surfing, respectively. They share a heritage of starting off life as responses to functional needs and have been deemed ugly or unattractive. These brands have earned the status of being known as "iconic," as they have evolved from utility items to high-fashion commodities, finally settling as staple "must have" items. In order to continue their evolution beyond their original intended purpose and to become sustainable businesses in the fickle world of fashion, they often diversify via new product categories and markets.

8.9

8.9 Alternative use for the iconic Hunter Wellington boot
A treasured possession is recycled and still retains a practical purpose true to its heritage.

ETHICS IN ACTION:
Social Enterprise

Brothers Rob and Paul Forkan founded their flip-flop brand Gandys with the vision of creating a sustainable social enterprise that would help but also inspire others to create change. They lost their parents tragically in the Boxing Day Tsunami in 2004, and thus are committed to supporting children in need of basic essentials such as nutrition, shelter, and education.

The "Orphans for Orphans" mission has already started to make an impact by funding children's homes in India and Sri Lanka, and the vision is to open them all around the world. Gandys is a social enterprise and was founded on a commitment to use a portion of all profits to help disadvantaged children through a registered charity the Gandys Foundation (10 percent of all Gandys profits go directly into the Gandys Foundation).

8.10

8.10 Rob and Paul Forkan, Gandys Kids Campus, Sri Lanka
The "Orphans for Orphans" initiative facility encompasses a broad range of activities including academic support, woodwork lessons, and sports activities.

8.11

8.11 Gandys map flip-flops
For this authentic travel brand with a fascinating and unique heritage, Gandys "map" product conveys the brand ethos effectively.

" I ASKED MY BROTHER PAUL TO LEAVE AUSTRALIA A FEW YEARS AGO TO COME BACK AND HELP WITH AN IDEA TO DESIGN FLIP-FLOPS THAT WOULD ALLOW US TO BUILD A CHILDREN'S HOME IN MEMORY OF OUR PARENTS, WHO WE LOST IN THE BOXING DAY TSUNAMI. AFTER ALL THE TRAVELLING AND VOLUNTEERING AND THE SACRIFICE OF THEIR LIVES THEY MADE TO SAVE OUR YOUNGEST SIBLINGS, WE HAVE WORKED DAY AND NIGHT TO MAKE THIS IDEA BECOME A REALITY. THE PEOPLE OF SRI LANKA ALSO HELPED US GET TO SAFETY; WE WERE ONLY KIDS AT THE TIME SO THIS IS WHY WE HAVE RETURNED TO OPEN THE FIRST GANDYS KIDS CAMPUS, WHICH WILL HOPEFULLY SEE OTHER CHILDREN IN TEN YEARS' TIME REACH THEIR OWN DREAMS."
ROB FORKAN

10 P's Model

The 10 P's model outlined by Kincade and Gibson (2010) can be used to evaluate how the British international brand Gandys operates in what could be considered an already overpopulated single product category, flip-flops, engaging with its consumers on an emotional level and presenting itself in such a way that reinforces its own brand values.

1. **People**—target consumers are students and travelers.
2. **Product**—100 percent FSC-certified rubber, made in Fuzhou, China.
3. **Pricing**—entry price £18 ($30); ceiling price £30 ($50).
4. **Position**—affordable and unique.
5. **Placement**—from premium UK department stores (Selfridges and Liberty) to high-street multiples (Office and Sole Trader) and global e-tailer ASOS.
6. **Presentation**—merchandised in pairs on the retail floor, as per the market leader in this category.
7. **Promotion**—price promotion via its transactional website; four pairs for the price of three. A product collaboration with Liberty department store featuring Liberty classic paisley art printed insoles. Window, instore, and website promotions.
8. **Packaging**—re-cyclable.
9. **Processing**—product is shipped via boat, not air freighted.
10. **Playback (feedback)**—via its student brand ambassadors program and "Flip-Flop Friday" pitches.

CASE STUDY:
Dr. Martens

Following a skiing accident in 1945, Dr. Klaus Maertens, a German orthopaedic doctor, designed a shoe with an air-cushioned sole. He went into partnership with a friend and colleague, Dr. Herbert Funck, and together they perfected and patented the first design. The initial prototypes were fashioned from bits of old army uniforms, but the real innovation was the heat-sealing process they developed that joined the sole to the upper, thereby forming a series of strong, air-cushioned compartments. In 1947 the shoes went into production, handmade in Seeshaupt, Germany. In 1952 production moved to a factory in Munich to meet growing demand, and in the 1950s the vast majority of the sales (80 percent) were to women over 40, far removed from the antiestablishment image the brand garnered in later decades. In the 1950s it was a successful German brand, selling mainly to the domestic market, with a catalogue of 200 styles.

So how did Dr. Maertens become a "British" brand called Dr. Martens? In 1959, the company advertised its brand and new technology in an overseas trade paper (the UK journal *Shoe and Leather News*) and this new innovation caught the attention of established British shoe manufacturer Bill Griggs, based in Northampton. The exclusive UK license was secured by the Griggs brothers and, after several adaptations in design and marketing for the UK market, on 1 April 1960, the first classic eight-hole boot in cherry leather with yellow stitching was born—the 1460, as it was later christened.

Initially these boots and shoes were marketed as workwear commodities and rebranded as Dr. Martens, featuring the word "Air Wair" in italic script, said to be based on Bill Griggs's original doodles, bearing the tagline "with bouncing soles." The black and yellow graphics of these early advertisements, ribbon loop, and yellow stitching of the early models are testaments to Bill Griggs's eye for branding and have remained as a key design details today.

Initially the 1460, a workwear style, was worn by postal workers and the police, ironic given the antiestablishment wearers that would follow. As a reaction against the feminine styles of the "flower power" era in the mid–late 1960s, military wear became a popular choice, worn by young people for the same reasons as the original consumers, i.e., price, comfort, and durability. If "flower power" was a style adopted by those who could afford international travel and an alternative lifestyle, early adopters of the Dr. Martens boot were working class and adopted military clothing and accessories as an expression of dissatisfaction with the status quo, an expression of their sexuality and political beliefs.

Across five decades, Dr. Martens have been worn by skinheads, punks, psychobillies, indie kids, goths, travelers, school kids, and construction workers alike. What they share with other iconic brands is that the original marketing was not aimed at the eventual consumer. Brands such as Converse and Dr. Scholl also share this history.

The wearers of Dr. Martens often develop an emotional attachment to their beloved boots, or shoes, which goes way beyond an appreciation of quality footwear. In *Dr. Martens: The Story of an Icon* (Roach 2003) the D.J. Steve Lamacq describes his attachment to his favorite boots:

> "DM's are the perfect lived-in footwear, an extended part of your soul, really. I wore my oldest pair to Reading festival in 1997 and they were covered in mud. I thought to myself, 'This is where we have to split up, finally, after all these years.' So when I got back to my hotel room, I held a small funeral service, took some Polaroids of the boots, before I carefully placed them in the bin."

These inanimate objects have been afforded a higher status, as if themselves human, or a part of one's very being. This personalized appeal is intrinsic in the brand's longevity.

Dr. Martens continues to have a loyal following today that spans generations. It is a rare example of a brand retaining its credibility with the youth market, as well as the original early adopters simultaneously. These uncontrived roots have added to its appeal over time and not just one, but numerous "style tribes" rebelling against the mainstream have championed the brand. It became the ultimate symbol of counter-culture in tough times, although its image today could be viewed as more sophisticated and controlled.

8.12

8.12 Wycombe marsh mob, 1981 (photograph by Gavin Watson)
Dr. Martens were adopted by punks in the 1970s; they were practical and durable in times of austerity and came to represent rebellion and nonconformity.

A continued focus on product quality and a segmented distribution strategy underpin the company's success today, as well as some well-targeted and savvy marketing campaigns. The brand is able to maintain sales at full price across the majority of its range by careful and strategic retail sales planning.

Keen to capitalize on the importance of provenance, some of the production returned to Northampton (from Asia) in 2004. Typically these are the higher-priced, more premium lines and collaborations. Collaborations occur with celebrities, designers, and traditional textile companies such as Stephen Walters and Sons Ltd., a woven fabric specialist originally established in 1720 in

CASE STUDY:
Dr. Martens

Spitalfields, London. This kind of industrial collaboration appears to be diametrically opposed to celebrity collaborations such as Agyness Deyn and Daisy Lowe; however, when we consider the broad and diverse customer base, the brand's reach is wide, and this broad appeal reinforces its status as an iconic brand.

The brand is still true to its roots in terms of product. One of its senior designers, Claire Russell, states, "We love breaking the boot down and playing with the various different components, different soles, different welts; there's a lot you can be experimenting with. And yet it still has to feel authentic; we always try to stay true to our roots rather than just slavishly following fashion."

8.14

8.13

8.13 "Stand for Something" autumn/winter 2014 campaign (photograph by James Pearson Howes)
Campaign imagery still conveys antiestablishment punk values expressed by Gen Y.

8.14 "Stand for Something" autumn/winter 2013 campaign (photograph by James Pearson Howes)
Campaign imagery designed to appeal to a different demographic who interpret the brand in a new way by styling it with tailoring.

This is also a sentiment echoed by the designer Joe Casely-Hayford when he was approached to collaborate with the brand. "Based on the idea that Dr. Martens transcends the superficiality of fashion, I made my boots inside out to highlight the consistent quality of this product and the importance of inner integrity."

Joe Casely-Hayford is one of the many designers who have collaborated with the brand over the years. Others include Orla Kiely, Ben de Lisi, Olivia Morris, and Red or Dead. Wayne Hemmingway (the co-founder of Red or Dead), states, "Dr. Martens do not reflect youth culture; they are a part of it. The boots are on the feet of countless youth movements, and it means so much. Even if a youngster hasn't yet come across Dr. Martens, show him or her these images (from *Dr. Martens: The Story of an Icon*), get them to soak up these tales, and they will immediately recognize the significance."

Many different demographics and psychographics elect to wear the brand fifty-two weeks a year, often regardless of contemporary fashion trends and even practical considerations such as temperature; such is the brand's consistent lure and iconic appeal.

" DR. MARTENS' PRODUCT AND COMMUNICATIONS ARE ALL ABOUT DEMONSTRATING THE AUTHENTIC HISTORY OF THE BRAND AND ITS BOOTS AND SHOES THROUGH ITS DESIGN, MUSIC, AND CULTURAL RELEVANCE. THIS WILL RESONATE WITH THE DIVERSE CONSUMERS WHO COME INTO THE BRAND, BY CHAMPIONING AND ENGAGING WITH INDIVIDUALS WHO 'STAND FOR SOMETHING.' THIS IS ALSO TO REINFORCE ITS UNIQUE POSITIONING, NOT ONLY WITH THE TRIBES WHO HAVE ALWAYS WORN THE BOOTS AND SHOES, BUT TO DRAW IN NEW CONSUMERS WITH PRODUCT, MESSAGING, AND IMAGERY THAT IS RELEVANT TO THEIR LIFESTYLES AND ATTITUDES."
SIMON JOBSON, GLOBAL MARKETING DIRECTOR AT AIRWAIR INTERNATIONAL LTD.—DR MARTENS

8.15

8.15 Agyness Deyn at Venice beach wearing collaboration product, 2014 (photograph by Moni Haworth)
Dr. Martens partner with Agyness Deyn on product collaborations and campaigns.

Industry Perspective:
Tracey Neuls, Designer

Tracey Neuls launched her label in 2000 after winning the New Generation Prize at London Fashion Week. Among her peers, she is known as the Designers' Designer: shunning fads and trends, her brand is carefully managed to appeal to a knowledgeable and discerning customer.

What was your first role in footwear?

I started my own company aged twenty-nine, working for ten years before that as clothing designer, and so all of that learning was moved across to shoes.

What training/qualifications/experience did you have for this role?

I trained in fashion design, no footwear courses existed in Canada or northern America at the time. Having made shoes since the age of nine, making heels out of toilet roles, it's something that I've always done. When I moved here I realized there was a specialist college—Cordwainers.

In your opinion, which fashion footwear brand is the most iconic and why?

Maybe not the most personal attachment from a design perspective, but Jimmy Choo. They are very different from what we do, but from a customer point of view they are the most well known. For me, it's the old houses that have built their names over time, like Ferragamo and Roger Vivier. They do not regurgitate what's gone before, but have built their names on credibility. It's all about the product for me.

Which brands have worked together on the most interesting and effective collaborations?

Louis Vuitton with Nina Saunders and Grayson Perry are really interesting. My interest is in the art and design world, not conventional artists, more about a space and how this works.

Which fashion footwear brand has evolved and adapted well to consumer demands?

Prada still has a hold, even though they are a multi-brand but they are still such a credible footwear brand. Also the work they have done with Rem Koolhaas is really interesting.

Which fashion footwear brand do you admire from a brand management perspective and why?

Trippen. They are avant-garde, but still wearable. Their design umbrella is still very clear and understandable, and I admire their consistency. For them it's all about quality and comfort. They own their own factories and have an amazing returns policy because of this, and this makes their customer service pretty difficult to beat!

What kind of training/qualifications/experience would you advise someone to get if he or she wants to work in fashion footwear?

This really depends on what aspect of the business you want to work in. Sometimes in colleges designers are prepared for a large corporate company, but this is not always where the opportunities are. Getting different experiences with different sizes of companies can teach you a lot.

What is the single most valuable piece of career advice anyone has given you?

Do what you are passionate about. But also don't wait for the "perfect job." Any job is worthwhile! Half the students I meet want to have their own business, but before you do this you need to get experience first.

8.16

8.16 Tracey Neuls East End store in London
Limited space at the Shoreditch store does not preclude highly creative and innovative brand presentation; Tracey Neuls collaborated with interior design company Faudet-Harrison in 2011 for the store opening and has worked with many artists, designers, and brands since, including Nina Saunders, James Rhodes, and Cutler and Gross.

SUMMARY

Footwear brands have the power to influence consumer recognition, awareness, perception, and ultimately sales. Successful footwear brands manage their business by appealing to many different demographics and psychographics simultaneously. This can be achieved by partnering with like-minded brands on collaborative projects. The partner brands may not necessarily be other footwear brands; however, the brands will share synergies as well as create new perceptions that change buying patterns and behaviors as a result.

Brand owners' motivations and vision will shape strategies and the overall marketing mix, and this may change over time as the brand evolves and has to respond to external and internal pressures to remain current and relevant. Brands that are deemed to be iconic earn this accolade by considering their marketing mix from a holistic perspective and altering their strategies as needed to respond to market and consumer demands.

DISCUSSION QUESTIONS

1. Which is the most iconic global footwear brand and why?

2. Which is the most iconic footwear brand in your market and why?

3. What are the advantages and disadvantages of the social enterprise model for footwear brands?

4. What is the most successful footwear brand collaboration and why?

5. What criteria would you use to measure this success?

6. Which footwear brand has the biggest image problem and why?

EXERCISES

• Which are the most successful and effective women's, men's, and kids' footwear collaborations in your local market? Focusing on these three brands, research and evaluate why they are successful and how this is measured/visible. What future recommendations would you make (from a marketing perspective) for each of the three brands?

• Select an established brand in your local market that is losing market share. By conducting both secondary and primary research, identify issues with regard to consumer awareness, perception, and loyalty and propose a suitable collaboration. Within your proposal consider advertising, product, distribution, and how you will track and evaluate the success of the partnership.

• By conducting primary research, identify a functional non-fashion brand that is starting to be adopted and worn as a fashion commodity. Explore the reasons for this: What types of consumers are adopting this brand and why? What kind of image are they wishing to portray? How would you describe them?

KEY TERMS

Brand audit—a detailed examination of a brand in its current state, to assess consumers' perception now; used to inform future strategy

Brand collaborations—where two or more brands work together to design and market product or product ranges that enhance their reputation and are part of a broader strategic aim

Brand DNA—a metaphor that describes the interaction of fundamental elements of the brand that lead to the evolution of the brand as a living organism

Brand equity—a brand's power derived from the consumer's recognition that it has earned over time, which translates into higher sales volume and higher profit margins versus its competition

Brand heat—positive consumer perception about a brand that creates desire and acts as a purchase trigger

Brand resonance—significance of a brand in the eyes of the consumer, by evoking an association or a strong emotion

Brand value—in the case of consumer product brands, this can be measured through customer loyalty and staff retention/recruitment

Brand value (financial definition)—this is considered to be the net current value of the estimated future cash flows attributable to the brand

Co-branding—a marketing partnership between two or more brands; can encompass different types of branding partnerships, such as product or sponsorship

Consumer loyalty—a tendency to favor one brand over others, based on attitudes or behaviours, due to satisfaction, convenience, performance, or familiarity with the brand

Consumer perception—how individuals form opinions about companies and the merchandise they offer via the purchases they make

Core business—the primary product area or activity that a company focuses on; often first founded upon

Demographics—studies of a population based on factors such as age, ethnicity, sex, economic status, level of education, income, and employment; used by brands in a situational analysis of a specific market to help inform further research and strategy

For profit—a business whose primary goal is making money (profit), as opposed to a nonprofit or social enterprise organization, which focuses on goals such as helping the community and only seeks to make enough profit in order to keep the organization operating

Iconic brand—the "go to" brand that is the customer's first choice; the most sought-after brand within the product category

Over-distribution—when a brand is sold through too may access points, resulting in brand equity being eroded

Psychographics—consumer research that focuses on opinions, beliefs, and preferences to build a picture of the target group's lifestyle; often used in conjunction with demographics-based data

Social enterprise—a business that trades to address social problems or improve communities or the environment; profits from selling goods or services are reinvested back into the business and the community

09

MARKETING COMMUNICATIONS

Learning Objectives

- Define marketing communications and the key tools used to promote fashion footwear.
- Analyze business strategy behind integrated marketing campaigns and discuss measurements for success.
- Identify the new forms of digital promotion and how to measure their effectiveness.
- Assess the issues relating to regulatory and cultural expectations for brands operating in a global and digital marketplace.

9.1 Cast from Sex and the City
The cast of Sex and the City at the 2008 world premiere of *Sex and the City: The Movie* in London's Leicester Square. Sarah Jessica in Alexander McQueen peep-toe shoes, Kim Cattrall in Gucci strappy sandals, Cynthia Nixon in Gucci peep-toe shoes (not seen), and Kristin Davis in Christian Louboutin peep-toe mini-platforms.

INTRODUCTION

Establishing and maintaining a fashion footwear business, brand, or service requires consideration of various marketing activities and how to blend them together. Although awareness of the customer, the product, pricing, and where the product is sold is essential, it is also important to know how to communicate and promote the brand.

A company must define its message, be clear about who it is aiming the message at, and know what is the most appropriate and effective communication channel to reach them, while also considering the fundamental principles of marketing: price, product, place, and promotion. As discussed in Chapter 1, consumers expect a seamless experience across every touch point they have with the company or brand. This chapter will address the key promotional tools used when marketing fashion footwear and consider the current fusion of traditional and digital approaches to delivering messages to the potential consumer. In addition, we will also consider the rise in online marketing through websites, m-commerce, email, and social media.

WHAT IS MARKETING COMMUNICATIONS?

Marketing communications is also referred to as the promotional element of the marketing mix—the aspect of marketing that covers communication with existing or potential customers as well as trade and industry partners. A company is required to develop a set of objectives and a budget and utilize several communications tools for an effective campaign. Objectives can be to increase sales through new or existing customers, to increase on- and off-line traffic (visits to stores), and to increase brand awareness and loyalty. The communications mix consists of a variety of tools or disciplines that can be put together to send a message to the consumer. The key traditional tools are listed below and will be discussed in more depth throughout this chapter, with a focus on digital developments toward the end of the chapter. Marketing communications must also consider the stakeholders—people who have an interest in a company. They can be investors, employees, and those who help to promote the business, such as journalists and bloggers.

Advertising is paid for by the retailer or brand, delivered to an audience via mass media, and attempts to persuade the viewer to buy the product. Ads are placed in traditional media channels such as TV, magazines, billboards, and transit space such as stations and airports.

Public relations is the development and maintenance of good relationships with different stakeholders of the brand, such as journalists, bloggers, and stylists. Publicity can be more effective than paid for advertising, but it is necessary to create a story/idea/point of difference or interest that has high perceived value and will engage readers and viewers.

" **MARKETING COMMUNICATIONS ARE THE TOOLS A COMPANY USES TO DELIVER A RANGE OF PROMOTIONAL MESSAGES TO ITS TARGET MARKETS."**
THE CHARTERED INSTITUTE OF MARKETING, 2009

Events management and sponsorship of people and events allows an opportunity to foster brand awareness and loyalty. Companies connect their brands with the potent emotional experience at big events or celebrities, which can elicit positive feelings.

Direct marketing or selling includes newsletters, catalogues, and images that are sent directly to the consumer from the brand, either by post or email, with an objective to inform or elicit a sale.

Retail marketing, sales promotion, point of purchase (POP), point of sale (POS), and personal selling are also part of the communications mix within the retail environment and have been discussed within visual merchandising in Chapter 6.

Integrated Marketing Communications

It is essential to coordinate a brand's marketing campaign across all communication channels; this is commonly known as integrated marketing communications (IMC). There has been a move away from sending out mass messages to a large homogenous market, instead focusing on niche marketing campaigns to target the individual consumer, which has proven to be more persuasive. It is vital that these campaigns both are cost effective and reach the target customer. IMC campaigns now use a mix of traditional and digital media. A message will be transmitted through a variety of tools from the communications mix. The internet offers the potential for relationship marketing, focusing on customer loyalty and retention, fostering interactivity and more accurate customer profiling due to the large amount of data it can collect through online activity.

For campaigns to be truly effective, they must use a mix of traditional and digital promotion and be part of a broader marketing strategy, which is in turn part of the wider company strategy, rather than an isolated notion.

Planning with the end in sight is crucial, and agencies will be retained or not, as the case may be, based on business results. Creating compelling, intriguing, and even traffic-stopping campaigns is critical to engaging with the consumer. The AIDA model can be used to better understand this, as explained in Chapter 6.

> **" [IMC IS] A COMPREHENSIVE PLAN THAT EVALUATES THE STRATEGIC ROLES OF A VARIETY OF COMMUNICATION DISCIPLINES AND COMBINES THESE DISCIPLINES TO PROVIDE CLARITY, CONSISTENCY, AND MAXIMUM COMMUNICATION IMPACT."**
> AMERICAN ASSOCIATION OF ADVERTISING AGENCIES, 2013

9.2

THE WOLVES
Spring / Summer 2015

Danielle Romeril
Georgia Hardinge
Isabel Garcia Gold Label
Joanne Stoker
Orla Kiely
Preen Line
Mini Preen
Rejina Pyo
Toga

9.2 A PR agency showroom
Emerging and niche fashion and footwear brands may choose to work with a PR agency to promote their seasonal collections.

MARKET RESEARCH

As described in Chapter 1, market research is crucial to developing an understanding of the consumer prior to launching a campaign. There are several months of research and planning, ranging from four months to a year, depending on the designer or the brand. If the designer/brand is struggling—if it is not gaining traction quickly enough or perhaps losing market share, as part of its initial research a brand may choose to conduct a full brand audit.

A full brand audit encompasses internal and external facets; the facets that relate to marketing and advertising are shown in Figure 9.3.

Davis (2009) suggests that by conducting assessments pre- and post-campaign, the effectiveness can be accurately measured. The following questions can form a solid basis for measuring brand impact via research, when identifying issues and dilemmas:

- **Brand influence**: Why do people "buy into" the brand? What status does a brand have among a certain age group? For example, adidas formally partnering with Kanye West to enhance its credibility with Gen Y fashion consumers.

- **Innovations**: Do people consider any particular product or service as innovative? For example, Danish brand Roccamore makes fashion high-heeled product with a constructed orthopedic insole.
- **Brand communications**: How is the brand being communicated? Do people remember the message or call to action? What do people think are the brands values? For example, Swedish Hasbeens spring/summer 2015 "Go Natural" campaign; Helmut Newton–inspired photography depicting naked models wearing new-season styles connotes empowerment and promotes the brand's "slow fashion" message.
- **Sustainability**: Does the brand have an underpinning sustainability strategy that looks at social and environmental impacts? For example; the Gandys flip-flop brand raises money to build children's homes via its charitable foundation that supports their "Orphans for Orphans" initiative.
- **Differentiation**: Can people recall the brand over competitors? Do people confuse the brand with its competition? For example, intellectual property protection allows brands to protect their point of difference in the marketplace such as Timberland® and its iconic Yellow Boot™.
- **Online**: Does the brand have a strong online presence? For example, Jimmy Choo has a very strong online presence through blogs such as "Wearing It Today" by Laura Fantacci and "The Shoe Snob" by Justin Fitzpatrick.

9.3 Brand and audit considerations
Internal factors are the brand elements that can be controlled and managed from within via research and strategic implementation, and the external factors are the vehicles for change visible to the consumer.

9.3

Brand audit considerations

Internal	External
Positioning	Expressed via corporate identity—e.g., logos and other branded elements in relation to the competition
Brand values	Expressed via collateral—brochures, print materials, and trade show displays
Unique selling proposition (USP)	Expressed by advertising, sponsorship, affiliations, civic involvement
Brand promise, or brand essence	Expressed via the website, outlining service proposition and content marketing such as "lifestyle" blogs
Corporate identity and brand standards	Conveyed by ambassadors and stakeholders of the brand
Product innovation	Conveyed via social media, PR testimonials, and other assets—e.g., blogs, white papers, case studies, videos, articles, and books

- **Leadership**: Do any particular personalities behind the brand stand out? Is the brand based around a particular personality? For example, eponymous brands are often the expression of the brand owners themselves, such as Finnish brand Minna Parikka. With publicly traded companies, this is also especially important as it affects internal and external confidence and perceptions.
- **Internal measures**: What is the internal understanding of the brand values? If this understanding is not aligned internally, a brand will seek the services of an external agency to research and propose solutions. By employing a third party, an objective approach can be adopted to yield honest and insightful findings that can feed meaningfully into future strategy.

These are all important factors when considering a full "rebrand"; when planning campaigns, the marketing team can investigate and test some of these facets of the brand's business via surveys, focus groups, product testing, and in-depth business analytics, to arrive at a logical, informed set of goals that the creative brief will be based upon.

Measuring Success

Promotional campaigns, especially those that include advertising, are costly exercises and so measuring the return on investment (ROI) post-campaign is key. ROI is a measure of the profit earned from each "investment," or promotional tool used, and can be calculated as (Return – Investment) / Investment.

ROI calculations for marketing campaigns can be complex, due to having variables on both the profit side and the investment (cost) side. Some marketers believe that analyzing ROI per campaign is not truly reflective of a campaign's full potential, as this can be too restrictive by viewing a single campaign in isolation, which is not how the consumer may view it.

It is prudent to consider both quantitative and qualitative measures pre- and post-campaign.

Quantitative measures:

- Increase in footfall/online visitors, e.g., expressed as a number or a percentage increase.
- Increase in average spend, also known as average basket (or ATV—average transaction value). This can be expressed as an increase in sales value or as a percentage.
- Increase in conversion (often expressed as a percentage); applies to retail and online.
- Increased sales within a specific category or subcategory, often linked with a strategic goal.
- Increased market share, within the product category; expressed as a percentage.
- Increased marketing reach, i.e., number of consumers touched by the campaign, e.g., in thousands or millions.

Qualitative measures:

- Increase in brand recognition/awareness, often measured by primary research (questionnaires and focus groups).
- Increase in perception of brand image, often measured by a shift in the consumer perception of who the competition is, and value for money (VFM) or perceived value.
- The successful introduction of a sub-brand. This can be deemed as being successful by increasing the brand portfolio's customer base (overall) rather than cannibalizing numbers.
- Re-positioning of an existing brand. Success can be measured using both qualitative and quantitative measures.

9.4 Market research groups
Focus groups are valuable exercises in gaining honest feedback from either consumers or opinion leaders; they are informal, small discussion groups professionally facilitated and moderated to produce qualitative data (preferences and beliefs) on a specific product or campaign (see interview in Chapter 1 with Jason Fulton).

ADVERTISING

Advertising campaigns are the most expensive promotional tool and crucially are paid for by the brand. They can be created in-house or externally, via advertising agencies. Agencies can be employed for a single seasonal campaign or a longer-term fixed contract as part of an ongoing business relationship that will still be dependent on results. Creating impactful and commercially successful advertising campaigns is the number one priority for most marketing teams. Increasingly, the remit of advertising agencies has become more diverse, but creating successful campaigns that can be tracked and measured is still of paramount importance. In 2014, the top U.S. TV advertisers in apparel and footwear were dominated by the big sportswear brands, led by adidas, which echoes the dominance these brands have in footwear sales.

Magazines

Traditional print magazines and newspapers serve two main purposes: as a medium to reach a large target audience with advertisements and a method to report fashion news and features through editorial. Magazine revenue is generated by their advertising sales, but their credibility and success is formed by their journalists and editorial content. Their ability to set the appropriate tone of voice and offer insight, inspiration, and style advice to the target reader is leveraged against their advertising rates. Leading women's fashion magazines such as *Elle*, *Harper's Bazaar*, *In Style*, *Marie Claire*, and *Vogue* frequently feature footwear editorial with titles such as "60 shoes to wear all spring" or "30 must-have ankle boots." Footwear may also be called in by journalists to use in photo shoots. Editorial is driven by newness, and often small, innovative, and edgy brands add variety to the pages. However, it has long been noted in the industry that editorial content is heavily influenced by those that advertise in the magazine.

As an advertiser, choice of magazine will depend on the circulation figures and readership, which is available in the media pack. The position of the ad will depend primarily on affordability. The nearer to the front of the magazine, the greater the cost to advertise. The fall and spring issues of fashion magazines are heavily weighted with ads, as this is when brands traditionally launch their new collections to the consumer.

The print magazine market has been steadily contracting with the availability of hardware such as tablets and smartphones, as well as the increase of free digital content. In the UK, sales of magazines have decreased by 36 percent since 2009. Consumers' desire to disconnect with technology is cited as a key reason to buy a magazine, but decreasing sales figures and free digital content will not halt the decline.

" IF YOU EVER HAVE THE GOOD FORTUNE TO CREATE A GREAT ADVERTISING CAMPAIGN, YOU WILL SOON SEE ANOTHER AGENCY STEAL IT. THIS IS IRRITATING, BUT DON'T LET IT WORRY YOU; NOBODY HAS EVER BUILT A BRAND BY IMITATING SOMEBODY ELSE'S ADVERTISING."
DAVID OGILVY

" BY ADVERTISING IN A GIVEN MAGAZINE, A COMPANY OBTAINS A DIRECT BENEFIT IN TERMS OF PRODUCT OVERAGE IN THE PAGES OF THAT MAGAZINE. HOWEVER, THE COMPANY ALSO OBTAINS AN INDIRECT BENEFIT BECAUSE COMPETING MAGAZINES WILL BE PRESSURED TO COVER THAT COMPANY'S PRODUCTS, EVEN WHEN THAT COMPANY MAY NOT BE AN ADVERTISER."
RINALLO & BASUROY, 2009

IFC DPS* £ 54,963	IPage front half £ 18,321	All non-specified DPS sites pro rata.
1st DPS £ 48,770	Page beauty £ 18,321	Further loadings will apply to guaranteed RH and/or consecutive pages
2nd/3rd DPS £ 44,748	Page ROM £ 14,800	Loose inserts from £45 per 000
Editor's letter £ 24,385	DPS ROM £ 33,312	Bound-in inserts from £76 per 000
The Look opener £ 23,276	IBC £ 22,168	Hand inserts from £105 per 000
What's Now £ 22,168	OBC £ 27,481	*Prices of gatefolds on application

Banner/Leaderboard	£25
MPU	£23
Double MPU	£30
Portrait	£40
Billboard	£35
Side Skins	£45
Page takeover*	£120
Solus Newsletter+£500 prodiction	£120
60/40 Newsletter+£500 production	£120
Standard Advertorial	£2,000
Look book from MPU	£36

*Page Takeover - Includes MPU or DMPU, LB or Billboard & side skins. No longer than a week on homepage

9.5–9.6 Rate card from InStyle magazine
The media pack and rate card outline key readership and circulation statistics that should fit with the brand's customer profile and marketing strategy. Detailed costs and deadlines for adverting images and content also allow for time and budget planning for a brand.

9.5

9.6

PUBLIC RELATIONS

To understand the role of public relations (PR) in fashion footwear, the definition of PR below is a succinct starting point, as it highlights that a company's words as well as actions are critical. Unlike advertising, whereby media exposure is gained as a result of payment for commodities such as page space or airtime, PR operates via third-party endorsement and carefully orchestrated reputation management that is either run from within the company in-house, or externally via an agency. Agencies are employed by a designer or a brand, whereas retailers with private-label (own-label) brands often manage their PR in-house. PR companies are most commonly employed via a contract that specifies targets such as numbers of articles and photographic features within targeted publications, or TV appearances on particular shows that appeal to the target customer.

Agencies tend to work in three ways:
- a retained fee basis where they charge for a specified amount of time spent and activities undertaken, usually per month;
- a results basis, which is more akin to piecework whereby they invoice for an agreed output after this has been confirmed, such as an editorial with a specific publication;
- a project basis, whereby they charge upon completion for an agreed-upon set of multiple outcomes, such as introductions to key retailers and placement in editorial features.

" **PUBLIC RELATIONS IS ALL ABOUT REPUTATION. IT'S THE RESULT OF WHAT YOU DO, WHAT YOU SAY, AND WHAT OTHERS SAY ABOUT YOU. IT IS USED TO GAIN TRUST AND UNDERSTANDING BETWEEN AN ORGANIZATION AND ITS VARIOUS PUBLICS."**
PUBLIC RELATIONS CONSULTANTS ASSOCIATION (PRCA)

9.7

9.8

9.7 Joanne Stoker spring/summer 2015 press release and extract from forty-page look book

Joanne's inspiration for her spring/summer 2015 collection was 1969. The press release and look book are designed for journalists and bloggers to refer to when writing about and calling in samples for photo shoots. They will keep a selection of information from presentations and press days they have attended during seasonal fashion weeks.

9.8 Invitation to Joanne Stoker spring/summer 2016 digital presentation at London Fashion Week

Joanne's inspiration for her spring/summer 2016 collection was 1981. She used recycled records found in thrift stores and charity shops and reworked them as personalized invitations to fashion press and buyers. The event included a digital film, models, and dance presentation from House of Voga. This allowed press to view the full collection of footwear and accessories.

Press Releases

A press release is a statement prepared by the PR team for distribution to the media. The purpose of a press release is to give journalists information that is useful, accurate, and interesting. Press releases should conform to an established format, as journalists and bloggers have set standards and expectations. Companies should also offer interactive press kits made available online where information and images are available to download.

When writing a press release, the following should be considered: who, what, where, when, and why.
- The first paragraph of the press release should contain in brief detail what the press release is about—who and what.
- The second paragraph explains in detail: who cares; why you should care; where one can find or see it; when it will happen; product information—where, when, and why.
- If there is a third paragraph, it contains a summary and/or further company/contact info.

The aim of a footwear press release is to see the shoes featured in magazine or TV editorial. Brands may highlight "hero" or press pieces that are more eye catching and interesting than more commercial styles. However, it is not easy to assess the financial benefits of hero styles, as sales of these specific styles can be quite low; however, a brand "halo" effect can be created, enhancing brand awareness and recognition and increasing overall sales. PR agencies collect the editorial coverage through a press cuttings service, and brands need to consider the publication, its circulation figures, and the journalist/stylist who is responsible and the synergy with the target audience.

Examples of PR Agencies and Brand Partners

Surgery PR, London—Birkenstock, Superga, Frye, Sebago, Minnetonka, Blunstone, and Seven Boot Lane
Starworks Group, New York, Los Angeles, and London— Hunter, Jimmy Choo, Nicholas Kirkwood, Robert Clergerie, and Sophia Webster

9.9 Haviana press examples from flagship store, New York City
Current fashion publications are placed in prominent locations on the shop floor to raise awareness of advertised styles, or "hero" styles; the consumer is encouraged to connect the campaign to retail reality and make a purchase.

9.10 Oka-B press cuttings
Examples of press cuttings, which are a useful wholesale tool, especially when presenting to retail buyers.

Product Seeding

Another function of the PR team is to ensure that product is seen on the right celebrities and upcoming "undiscovered" talent in the creative industries. This activity is known as "product seeding," whereby these artists, musicians, writers, bloggers, actors, or models are gifted product, via the PR teams, on behalf of the brand or designer.

PR teams plan these celebrity targets by maintaining and updating lists of influential individuals who are considered innovators or early adopters (see Rogers' diffusion model in Chapter 1). The PR teams cultivate relationships with these individuals and their management teams and try to ensure that the individuals are photographed in the product. This process can be difficult to manage and control, although it is considered highly effective, especially in the digital age, and when successfully achieved can be beneficial to both the brand and the celebrity by reinforcing their respective identities by association.

These lists can change quickly and are often the source of debate within brands themselves, because the choice of celebrity is crucial in how the potential consumer views the brand; as such, the choice of celebrity/celebrities needs to directly reflect the brand's identity and values (see brand identity in Chapter 7).

THE STYLIST'S ROLE

The importance of the celebrity stylist, as the conduit from designer to celebrity, has grown rapidly since the late 1980s, where top-earning stylists can earn tens of thousands of dollars per job. With multiple clients and on the product seeding lists of PR teams themselves, these "friends of the brand" wield increasing power. Cristina Ehrlich, Erin Walsh, Leslie Fremar, Kate Young, Elizabeth Stewart, and Johnny Wujek are some of

the most influential figures in celebrity styling. In this symbiotic relationship, the celebrity looks glamorous while outsourcing the responsibility of creating and shopping for their "look" to a tried, tested, and trusted source, and in return the brands get images disseminated in nearly real time across the web. Designers and brands view this as a sound investment given the difference between more traditional marketing methods such as costly press advertisements, when the power of celebrity is so effective that items can become instant sell-outs with company websites even crashing on occasion due to the demand that is created.

9.11 Fashion blogger Elvira Abasova wears Kat Maconie
Elvira Abasova, co-founder of the "The Russian Code" fashion and beauty blog, attends Paris and Milan Fashion Weeks and personifies Kat Maconie brand values of glamour and luxury.

CASE STUDY:
Courting Hollywood

Product seeding is by no means a modern phenomenon. In 1923 Salvatore Ferragamo opened the Hollywood Boot Shop and made shoes for stars such as Joan Crawford and Gloria Swanson, as well as for films such as Cecil B. DeMille's *The Ten Commandments* (after emigrating from southern Italy to Boston and then California in 1914). In 1927 he returned to Italy and set up a shoe shop in Florence. His early days in Hollywood taught him the importance of the role of celebrities in marketing his evolving shoe business, and by working with stars such as Greta Garbo, Audrey Hepburn, and Sophia Loren he was able to promote his designs via the feet of Hollywood's most famous and glamorous.

The first red carpet dates as far back as 458 B.C.E. to the Aeschylus play *Agamemnon*, which depicts a Trojan War hero who returns home to find a crimson carpet rolled out for him by his wife. It is said that the late Hollywood showman Sid Grauman may have created the town's red-carpet tradition by creating a crimson-colored walkway in front of his Egyptian Theatre for the first-ever Hollywood premiere, *Robin Hood* starring Douglas Fairbanks, in 1922.

Since the 1920s, the tradition of the red carpet has continued to evolve, from the first Academy Awards ceremony in 1929, the Oscars, and the first Academy of Television Arts & Sciences awards in 1949, the Emmys.

The tradition of the red carpet has become increasingly influential as its own entity. No longer the sacred runway where studio-groomed stars could be admired from a distance; with the advent of TV shows such as *Live from the Red Carpet*, co-hosted by the late Joan Rivers and her daughter Melissa, the duo revolutionized the red carpet by entering into a dialogue by asking "Who are you wearing?"

From the late 1980s onwards, designers were keen to foster close working relationships with Hollywood stars and their management teams, before the notion of stylists became commonplace at such events. In 1988 Giorgio Armani set up his Rodeo Drive store and began courting Hollywood actors in the '90s, and outfitting them on the most important nights of their careers, the award ceremonies. He dressed Jodie Foster, Michelle Pfeiffer, Julia Roberts, Tom Hanks, Denzel Washington, and Billy Crystal. Soon other fashion houses followed suit, making Hollywood and Los Angeles essential locations for retail stores and PR offices.

Jimmy Choo, Christian Louboutin, Manolo Blahnik, Stuart Weitzman, and Calvin Klein use the red carpet as their stage and are consistent favorites of the stars, with brands such as Charlotte Olympia and Sophia Webster gaining popularity.

9.12

9.13

9.12 Katy Perry
Katy Perry wears gold Jimmy Choo "Lauren" sandals and a dress by Marchesa at the launch of *Katy Perry: Part of Me* at Empire Leicester Square, 2012.

9.13 Susan Sarandon
Susan Sarandon wears pointed black flat Jimmy Choo "Lucy" shoes and a black tailored Saint Laurent suit to the premiere of Woody Allen's film *Café Society* at the 69th Annual Cannes Film Festival.

EVENTS MANAGEMENT AND SPONSORSHIP

Managing events on behalf of a designer or brand is part of the public relations (PR) remit. These events can be open to the public (business to consumer, or B2C) or for trade only (business to business, or B2B). The desired outcomes of these events should be part of a wider strategic plan, rather than merely parties or happenings, with SMART goals that are set prior to and evaluated after the event.

Sponsorship

Rather than hosting an event, brands may decide to sponsor an event. This can achieve some of the outcomes outlined in Figure 9.12 but without the same commitment of resources such as staff and finance. Sponsorship can furnish a unique opportunity to foster brand loyalty and increase awareness, as brands can connect with the potent emotional experience at big events. The brand (sponsor) provides finance to partly fund an event and acquires the rights to display a brand name, logo, or advertising message on-site. This may also include a pop-up shop, exhibition space, and VIP product seeding areas (for example, Hunter sponsored the Coachella music festival in 2015) or the opportunity to retail at a sporting event such as the Badminton Horse Trials.

9.14 Types of events and desired measurable outcomes
Events are costly and require planning, with clearly defined measurable outcomes that can be targeted and tracked.

9.14

	Type of event	Desired measurable outcomes
Business to business (B2B) **Also known as trade marketing (discussed later in this chapter)**	Fashion shows trade fairs	Uplift in forward orders from retail buyers versus previous year
	Press days	Increased number of editorials and product placement by journalists and bloggers
	Charity fundraisers	Achieving a predetermined target for the charity and part of wider CSR plan
	Seasonal launch parties	Develop network, brand awareness, and increased number of buyers; contact details captured and sales appointments secured
Business to consumer (B2C)	Fashion shows	Increased brand recognition and awareness and sales uplift in local mall/retail outlets
	Pop-up stores	Additional sales revenue and opportunity to clear stock and samples
	Retail openings	Increased sales revenue and increased brand recognition and awareness
	Sports or lifestyle events such as festivals	Additional sales revenue and opportunity to clear stock and samples and increase brand recognition and awareness

DIRECT SALES

Linking directly to the end consumer to elicit a sale is also considered a promotional tool. Traditional strategies through home catalogues are still a core direct marketing activity; however, the internet is now a necessary line of communication, as discussed later in this chapter. Direct marketing is especially important in a crowded marketplace and where footwear can be considered a high-risk purchase due to issues of fit.

Trunk Shows

A trunk show is a private event, usually held at a retail store in collaboration with the brand or designer where guests view an edited selection of the collection ahead of season. The term was originally coined because sales representatives during the nineteenth century would travel across the country with their wares in trunks.

9.15 Joules lifestyle brand at Badminton Horse Trials
Retailing at busy sporting events such as Badminton Horse Trials can be lucrative as well as an excellent way to connect with consumers.

It is an opportunity for both store staff and clients to view merchandise ahead of season (typically three to four months before it lands, or four to eight weeks if it's a short-order brand) and to sell prototypes and previous season samples. These events can be advertised and open to the public, or run by invite only, depending on the brand/designer or the retailer.

The advantages are that they allow representatives from designers/brands to present in-depth information on product for the upcoming season, as well as be there to answer any technical and styling questions that the store staff and clients may have. These are often sociable and glamorous occasions and provide good opportunities for clients new to a brand to purchase a bargain on the day, or to feel more confident about preordering for the upcoming season. From a practical perspective, this gives both the retailer and the brand/designer an early read on what the best-sellers may be within season, and for independent boutiques they can commit to a new line knowing that there are customers for this new designer/brand that they are testing. Payment is taken by the retailer in full on the day, or a 50 percent deposit is placed with the remaining 50 percent paid when the goods are collected from store.

From a marketing perspective, valuable insights can be gained from these events, akin to a focus group. Designers and company representatives can glean valuable firsthand insights into who their customers are and what their tastes and preferences are. These events are starting to gain traction in the UK, where they have not previously been a feature on the retail calendar; this is especially relevant after the 2008 financial crisis, with many brands having to canvass and listen to customer feedback more than ever before.

Trunk shows are especially popular with independent boutiques and designer areas within department stores; Saks Fifth Avenue ran an event "The Pursuit of the Perfect Pair" with Oliver Sweeney allowing its best clients to preorder styles before they were made available to the general public, as well as being able to order custom-made styles.

There are some challenges for footwear with trunk shows. Typically, samples are only made in one or two sizes, usually a U.S. size 6 or 7 (a UK size 4 or 5). If a brand/designer is able to carry a full-size run, this is advantageous, although this is not always possible or practical. However, if this is a brand that the retailer and customer is familiar with, and the brand has a good history with quality and sizing, this does not present too much of an issue.

Interestingly, a relatively new trend in this concept is now the "virtual trunk show." In December 2000, Stuart Weitzman hosted one of the first-ever virtual trunk shows, allowing customers to access their pick of fifty unique styles from the spring 2001 collection. The brand offered sizes from width AAAA to C, sizes 4 to 12; customers were able to preorder their exact size, thus offering an additional

9.16

service and allowing the brand to plan production more easily ahead of season. In this respect, the retailer is cut out of the loop, and the traditional notion of a trunk show has evolved to be an effective marketing technique.

Sales Promotions

The purpose of footwear sales promotions either in store or online is to stimulate short-term demand and encourage brand switching. Flash sales such as "Black Friday," the last Friday in November, are annual discount promotions encouraging consumers to purchase for the December holiday period.

Promotions can highlight a new brand or brand extension, and if it is cheaper than a competitor, it could induce trial use. However, this can encourage customers to view the product through a price orientation, i.e., the customer only bought it because it was on sale, reduced, etc., and this creates challenges when building and maintaining a strong and credible brand image. However, if the desire is to move stock, create cash flow, and gain market share from a competitor, sales promotions can obtain immediate and often measurable results.

9.17

9.16 Sarah Flint trunk show at Tess and Carlos
Luxury New York City–based designer Sarah Flint partners with independent boutique "mini-chain" Tess and Carlos, who stock exclusive and up-and-coming European and American brands.

9.17 Discounted footwear at Discount Shoe Warehouse (DSW)
The use of discounting as a promotional strategy has increased in the last decade due to a wide choice of footwear and a savvy customer looking for a bargain.

SHOES ON SCREEN–DIGITAL DEVELOPMENTS

The traditional role of promotion or marketing communications has changed beyond recognition in the last decade due to the advent of digital communications. Previously, information about fashion styles and trends flowed one way—from the brand to the customer. The fashion brands, their publicists, and journalists controlled this information and traditionally disseminated it through the media of magazines. Information coming from consumers was mostly statistical, quantitative information, such as where the product was purchased and how much they paid for it.

The rise in internet access now means that today's consumers are geographically borderless; they are empowered by digital information, and their values, attitude, and lifestyles are more complex to unravel as they use different digital platforms to gain knowledge. Through combinations of video content, apps, and social media their motivations are more difficult to unpick because it is harder to gauge what is influencing them, online or off-line. Although a brand may retain tight control of its image through traditional media channels and store environment, to engage and foster loyalty with the consumer it must be perceived as honest and relatable through its social media channels. The internet allows consistent access to the consumer and is instantly updatable; this provides the "reach" (massive customer response potential) and "richness" (depth of information and extension to communications gives a quality of customer experience not previously available).

> **" WITH THE RISE OF SOCIAL MEDIA AND USER-GENERATED CONTENT SUCH AS BLOGS, THE BALANCE OF POWER BETWEEN FASHION PRODUCERS/INTERMEDIARIES AND CONSUMERS IN TERMS OF WHO SHAPES BRAND PERCEPTION AND FASHION KNOWLEDGE APPEARS TO BE SHIFTING."**
> CREWE, 2013

Websites

Websites are not only a vehicle to sell product, as discussed in previous chapters, but they are also a hub of information that a brand uses to communicate with its existing and potential customers. Content media in websites is equally as important as the transactional area and can increase the time a viewer spends on the site. People tend to spend very little time on one website, let alone one page, and a "bounce rate" measures the percentage of people who land on a page without navigating through. The average bounce rate is around 45 percent; any higher and a brand needs to assess why this is. In order to retain potential shoppers, websites are increasing their non-transactional pages.

Luxury and premium fashion e-tailers, such as shoescribe .com, are developing new and interesting ways to present their content via both transactional pages and "editorial" pages featuring guest editors, video interviews, and special editions. The lines between retail and editorial are becoming increasingly blurred as e-tailers devise more sophisticated ways to encourage customers to engage and stay on pages/sites for longer. This highlights the importance of content marketing across media platforms. "'Non-transactional" content is critical, not just at designer level; for example, Stevemadden.com offers the Steve Madden online magazine, providing styling advice, inspiration, and interviews with Steve Madden staff to attract and engage customers.

Email

Engagement through daily emails is essential for any company looking to foster a longer-term relationship with its target customers. Updates on new styles, editorial stories, and sales promotions are sent to those who have signed up for emails but are not necessarily existing customers. Emails can help drive traffic (potential customers) to the website and data capture (developing consumer profiles). It is important to consider the content of emails, as there is increased in-box competition all vying for the viewers' attention by using key words e.g., "insane deal," "big thanks," etc.

9.18

Mobile Devices

Around 31 percent of website views are accessed via a mobile device. With this figure set to increase, brands are moving to more responsive websites that adapt to the hardware the viewer is using. The increase of smartphone accessibility and consumers using mobiles to check email means the sites must render effectively on mobile as well as desktop. For example, the call to actions must to be clear and optimized for consumers clicking links with their fingers on mobile as opposed to a mouse on desktop. In addition, font size has to be readable without the consumer needing to zoom.

Video

Video content is increasingly important for marketers who use it to promote and demonstrate products, information, and ideas. Video can be lighthearted, attractive, and subversive, which can help develop viral campaigns or "word-of-mouse." Video channels and YouTube videos are being developed with annotation and the ability to add links to where product is available. In the increasingly technologically driven marketplace, the moving image will become ever-more important in delivering a message and driving sales.

In the Blogosphere

The concept of writing and posting pictures of one's personal opinion to a public audience was launched around the mid-1990s; by 1999 there were around fifty blogs. In 2005 AOL recorded around 8 million personal blogs, which rose to 184 million with 346 million readers in 2008. Platforms like blogger.com launched in the mid-1990s, meaning the writer did not need to code or build a website to have a web presence. By 2010, 2 million bloggers were listed as fashion industry on blogger.com.

While editorial features in print media are still considered the best PR, these opportunities are becoming eclipsed by other types of publicity. Opportunities to tap into fashion blogger audiences are rapidly increasing. Blogs can be categorized in several ways, such as a citizen blogger (who does not earn a living from blogging) or a professional blogger (collaboration with industry for financial remuneration), but it is clear that the lines between independent opinion, advertising, and editorial are becoming more opaque. Brands offer well-established bloggers with a large following sponsorship and advertising revenue, and it is sometimes the same writer or blogger who will work across his or her personal or independent blogs at the same time as representing corporate blogs.

> " **BLOGS ARE INTERNET SITES WHERE INDIVIDUALS POST THEIR THOUGHTS, IDEAS, AND INSPIRATIONS ONLINE IN AN UNEDITED AND SPONTANEOUS STYLE.**"
> CREWE, 2013

9.18 The rise of the fashion blogger
Fashion Blogger Sofie Valkiers wearing Christian Louboutin shoes at Paris Haute Couture Fashion Week, autumn/winter 2014.

Referred to as digital or brand influencers, fashion and beauty bloggers are gaining more credibility with the millennial generation, who are less influenced by the traditional celebrity endorsement. Bloggers are considered more authentic and accessible, as they are viewed as regular people who have used and tested the products that they write about (blogging) or talk about (vlogging). They are seen as trustworthy despite the fact that they will probably have been paid for the endorsement. Many brands are now considering how to incorporate this type of influencer marketing into their communications strategy.

Both the Federal Trade Commison (FTC) in the USA and the Advertising Standards Authority in the UK issue guidelines for bloggers. If they are paid to write a positive review they must declare this as advertising to avoid misleading people and breaking the law. Whatever the situation, PR representatives must be aware of key and upcoming bloggers who act as opinion formers and credible source disseminators of information. They drive traffic to websites and increase search engine optimization (SEO, discussed further later in the chapter).

Social Media

Social media is immediate and interactive, and social networking sites such as Facebook and Twitter have opened up a dialog with the consumer with information being available instantly for a relatively low cost. Choice of network and content will depend heavily on who the users are; for example, Twitter tends to engage with fashion professionals, while Facebook is B2C; Instagram is visual and micro-blogging sites such as Pinterest and Tumblr are creative platforms to collect ideas, images, and concepts both for brands and their customers. The choice of social network must be appropriate to the brand and target customer; however, it is important to remember social media is driven by the viewer who may not be someone who will actually purchase the brand but enjoys engaging and following. For that reason, it is often difficult to see a direct return on investment, but this activity creates content and conversation that other potential customers may see.

Chiara Ferragni Footwear

Italian Blogger Chiara Ferragni started her blog, "The Blonde Salad," in 2009 and is now an established fashion personality, having worked with various companies as a model and spokesperson. She has appeared on TV and in print media, most recently being the first blogger to appear on a Vogue cover (Spain, 2015). Her company now employs over ten people and has over 200,000 twitter followers, 1 million Facebook likes, and 3.5 million Instagram followers. Ferragni has leveraged this impressive rise to launch an eponymous Italian-made footwear line, suggesting the importance of an online profile to launch and sell footwear products. Established in 2010, the footwear line is best known for its glittery flats, and in 2016 was available both online and in over 300 international boutiques.

9.19

9.19 Chiara Ferragni footwear autumn/ winter 2015 collection
Fashion bloggers are using their influence and access to customers to expand into other areas in fashion, such as footwear design.

9.20

Advantages of social media:

- Brands can see who/what your customers are following and commenting on and how they behave online—helps to develop a profile.
- Social is fast and reactive to new trends in technology, society, and fashion.
- It is inexpensive compared to traditional advertising.

Disadvantages of social media:

- Democratic means the company must respond very quickly or lose control.
- Managing the process with the right staff may cost money.
- It is still a very new form of promotion and communication and can change very quickly.
- Brands must choose the right site and use it in the right way.

Social media is a tool that appears to sit across both online and off-line marketing. Unlike other digital marketing tools, many retailers do not see a direct effect on sales as a result of day-to-day social media campaigns. However, volume of followers and likes are an important reflection of a brand's popularity and are important in terms of reach. Social media content is much more relaxed when compared to other channels, specifically email. Social media posts aim to encourage engagement, to increase reach, and to therefore attract new members; sale-specific messages are therefore less important. Content is to be conversational and relevant— social media is as much a PR/branding tool as it is a digital marketing method.

CONNECT. FOLLOW. SHARE. TWEET.
SEE WHAT WE ARE DOING. SHOW US WHAT YOU ARE DOING.

9.21

9.20 Oka-B social media campaigns
Brands will use a variety of social media platforms to gain access to different types of customers. A consistent and coherent message across all channels is essential, but viewer responses may vary.

9.21 Instagram app and micro blogging site
Instagram is increasingly popular due to its highly visual content.

MANAGING AND MEASURING EFFECTIVENESS ONLINE

The main objective of e-commerce marketing is to drive traffic to a brand's website and for that visit to result in a sale. Traditional/off-line marketing tools aim to raise awareness, increase branding, and indirectly improve sales. E-commerce marketing is a highly trackable and measureable method of marketing. A customer's journey to and around the website gives insight to his or her shopping behavior and influences marketing strategy. The data acquired from customers shopping online provides a vast database of information that can be analyzed; customer profiles can be identified and marketing strategy implemented to target the consumers with relevant sales-driving messages.

Google Analytics

Google Analytics (GA) is a web analytics service that provides detailed information about the traffic (visitors) to a brand's website and their behavior once on the website. GA has tracking pixels built into the back end of the website and, combined with URLs, enable analysis of visitors and shoppers. GA holds data on where visitors to the website come from, how long they spend on the website, what pages they viewed, and if they made a purchase. It also provides information on which device a visitor was browsing/shopping on—desktop, mobile, tablet, and demographic characteristics—where shoppers live, their age, and other interest areas.

Search Engine Marketing and Search Engine Optimization

Search Engine Optimization is a free method of marketing and is highly competitive. E-tailers must write original copy with key words to increase web searches on organic listings in search engines such as Google and Bing. Retailers that stock many well-known brands must stand out from competitors stocking the same or similar product. From an SEO perspective, they may prioritize key words such as brand names or celebrities over product categories if their customer is brand aware; for example, a "Sarah Jessica Parker" search should bring up links to the shoe line that is available at Nordstrom.

Paid-for Advertising or Pay per Click

Brands may choose to place banner ads on another website or search engine such as Google (Google ad words). For every click from the search engine through to the site, the business pays the search engine a pre-agreed-on price based on a bidding system. It is essential to get the ad on the first page of search results and ideally in the first one to three spaces using key words that have been identified.

" **EIGHTY PERCENT OF POTENTIAL CUSTOMERS GO TO GOOGLE WITH A KEY WORD SEARCH—THEY KNOW WHAT THEY ARE LOOKING FOR—DO YOU?"**
GOOGLE

9.22 La Paire seasonal launch event at London Fashion Week
Displaying current season "show cards" at trade events helps to persuade retail buyers to try out new brands.

" I FEEL INCREDIBLY LUCKY TO BE RUNNING MY OWN BUSINESS AND CREATING A PRODUCT THAT I'M PASSIONATE ABOUT SO THERE ARE MANY THINGS THAT I ENJOY ABOUT THE JOB, BUT IT'S THE CREATIVE ASPECT THAT GIVES ME REAL JOY—FROM SHOE DESIGN TO CREATING, STYLING, AND PHOTOGRAPHING THE PRODUCTS FOR OUR MARKETING MATERIAL AND ONLINE CONTENT."
AMY LA, OWNER AND FOUNDER OF LA PAIRE

9.23

9.23 Yosi Samra spring/summer 2015 look book image
Yosi Samra maximizes sales at international trade shows by supplying future season look-book imagery to help create brand awareness.

TRADE MARKETING COMMUNICATIONS

Trade marketing is a complementary discipline to brand marketing. Its goal is to increase demand with supply chain partners, such as distributors, wholesalers, and retailers, rather than at consumer level. It is the process by which brands ensure that they are able to supply to meet the consumer demand created by brand marketing. This is a business-to-business activity.

Seasonal Launches at Trade Fairs and Showrooms

When brands are looking to expand locally and internationally, they select trade shows carefully to ensure they attract the attention of the retailers they are targeting. London has two main premium shows, London Fashion Week and the recently established show "Scoop," which since 2011 has become a key destination showcasing over 250 international contemporary designers. Upcoming British brand La Paire shows at London Fashion Week and Scoop.

"Trade" Marketing Targets in Fashion Footwear

Buyers—retail buyers representing department stores, multiples, boutiques, and pure-play e-tailers are the primary targets, as they hold the key to future sales.

Journalists for trade press—such as Drapers, Footwear Today, and Retail Week (in the UK) and Footwear News and Footwear Plus (in the United States).

Bloggers—these may be impartial but increasingly are paid directly or indirectly by brands.

Vloggers—video diarists who may also have a business agreement with PR agencies and brands. Designers, manufacturers, and suppliers—important opinion leaders within the trade due to their technical knowledge.

Competitors—established successful brands often have closed stands at trade shows, or separate invite-only showrooms away from the main trade show locations to protect the IP for the upcoming season; however, it's important for selected information to be made available to competitor brands ahead of season, and this will be tactically released information and imagery.

La Paire is an up-and-coming boutique shoe brand specializing in handmade sandals and flats, designed in London and made in Spain by traditional artisans. Establishing its premium positioning and point of difference in a competitive wholesale landscape has been achieved by showing at both shows. Generating her own PR and marketing content, the brand's owner and founder, Amy La, has quickly established some influential retail and press partnerships through these shows with Red magazine, which also has its own transactional fashion website called Red Direct, and Wolf and Badger, which not only supports up-and-coming young independent designers through its Mayfair, Notting Hill, and online stores, but also via its own multi-branded seasonal showroom at London Fashion week.

New York brand Yosi Samra also shows at Scoop, London, where it is able to increase its distribution from over 1,000 boutiques across the United States and eighty-five other countries, including fifteen brand shops in Asia and the Middle East. A second-generation shoe designer, Yosi Samra grew up in his father's factory in New York City and leads this growing brand, which first launched in 2009. Look books are an important tool for the company, as these images can be used by press and retailers alike to promote the brand.

ETHICS IN ACTION:
Regulatory and Cultural Considerations

The old advertising adage of "sex sells" is more relevant today than ever before. With brands seeking to expand into international markets, not only do they have intellectual property issues to explore prior to launching in new territories (see Chapter 7), but there are ethical and cultural factors to consider, as discussed in Chapter 5. Different countries operate varying systems for controlling advertising standards, and these can be monitored and enforced by one main regulatory authority.

- In the United States, the Federal Trade Commission (FTC) is a federal body; therefore, businesses of all sizes have a legal responsibility to ensure that any advertising claims are truthful, not deceptive, and that marketing activities do not break the law. The FTC oversees and regulates advertising and marketing law in the United States. (http://www.business.ftc.gov)
- In the UK, the Advertising Standards Authority (ASA) is independent of the UK government but is funded by the advertisers themselves via levees placed upon advertising space and some direct mail. This NGO (non-governmental organization) is reactive in that it operates in response to complaints from the general public, so in essence this is a self-governing system. It applies the Advertising Codes, which are written by the Committees of Advertising Practice. If it judges an advertisement to be in breach of the UK Advertising Codes, the ad is then withdrawn or amended, and the advertiser must not use the approach again. (http://www.asa.org.uk)
- In the United Arab Emirates (UAE), the National Media Council (NMC) is a federal government body that controls, monitors, and supports advertising and marketing within the territory. Its vision is to reach international standards in the field of media regulation to achieve the highest degree of transparency and excellence in the UAE; however, there are strict guidelines that global companies must adhere to, often having to amend and develop alternative advertising within the region. (http://www.nmc.gov.ae)

Cultural and religious issues are very important for brands to consider before they start selling and marketing in new territories. With some of the fastest-developing retail markets being in the Middle East, it is important that the media plan is cognisant of the new markets' traditions and laws.

The United Arab Emirates is one of the fastest-growing markets globally, with retail estimated to increase by 32.9 percent by 2015 from AED 114 billion in 2012 to AED 151 billion, according to Dubai Investment Forum's (FDI) latest report. Contributing factors include high disposable incomes, retail real estate expansion, and the continued construction of shopping malls and hypermarkets. Sales growth for the luxury goods market is expected to outperform global counterparts with a compound annual growth rate (CAGR) of +8.5 percent from 2010 to 2015, which has contributed to driving overall retail sector growth.

In turn, this retail growth has led to a significant increase in advertising spending, and companies such as the Alshaya group, which operates the Harvey Nichols franchise in Dubai (part of the UAE) must develop specific advertising that is in keeping with local cultural taste and religious considerations.

9.24

9.24 An advertisement for Harvey Nichols in Dubai
Bespoke brand imagery is tailored to each local market.

CASE STUDY:
Creation of a Heritage Brand—Hunter

Having earned iconic status by being true to its heritage since its inception in 1856, in 2013 the traditional Wellington boot brand Hunter appointed Alasdhair Willis as creative director; the brand has been reignited and become relevant to a new target audience, while still retaining its authenticity and credibility. At the start of his tenure, the brand staged a runway show at London Fashion Week for autumn/winter 2014. The purpose of this was to create a more contemporary, high-fashion profile.

Having capitalized on the "Kate Moss effect," when she first wore the brand at the Glastonbury music festival in 2005, the brand saw a resurgence in popularity as it was adopted by fashion consumers, rather than being donned for purely functional reasons.

Hunter now works with Starworks Group PR and Fashion PR, an agency that specializes in events such as fashion shows, press open days, and launches, as well as the Pencil agency, which describe itself as being "Couture Content Creators." Hunter continues to work with festivals such as Glastonbury, where it seeds product with celebrities such as Alexa Chung, Ellie Goulding, and Rita Ora. It also partners with the Port Eliot festival in Cornwall, an up-market arts festival that has a walled garden that every year morphs into The Wardrobe Department. This area was created by Sarah Mower (the British Fashion Council's ambassador for emerging talent) and features brands such as Hunter, Mary Katrantzou, Anna Sui, Barbara Hulanicki, and Luella Bartley.

9.25

Hunter is also one of the official partners and kit suppliers of the annual Oxford versus Cambridge University boat race, a traditional rowing event that first ran in 1829. As such, Hunter targets multiple demographics and psychographics via the strategic involvement by partnering with specialist PR agencies with expertise in events management.

> " HUNTER IS A BRITISH HERITAGE BRAND WITH A RICH HISTORY OF INNOVATION AND IS RENOWNED FOR ITS ICONIC ORIGINAL BOOT. ESTABLISHED IN 1856, THE BRAND IS WORN BY THOSE WHO LEAD, FORGING DESIGNS TO SHIELD PIONEERS FROM THE ELEMENTS AND THE BLEAKEST OF LANDSCAPES."
> HUNTER, 2015

9.25 Hunter boots are a festival staple
Hunter is the boot "brand of choice" for many festival-goers; the festival season has become so important to the brand that it is a major driver for both campaigns and product design.

9.26

1856
Founded in Scotland by American entrepreneur Henry Lee Norris; originally called North British Rubber Company, later to be known as Hunter

1914
The British War Office commissioned rubber boots to be worn in the trenches during World War I

1939
The North British Rubber Company produced protective rubber boots after the outbreak of World War II

1956
Marks the creation of the Original Green Wellington; the style later became known as the Original boot

1977
Awarded a Royal Warrant by Appointment to His Royal Highness Duke of Edinburgh

1981
Worn by Lady Diana Spencer in her engagement photos

1986
Awarded a Royal Warrant by Appointment to Her Majesty The Queen

2005
Kate Moss photographed at Glastonbury festival wearing Hunter

2006
The Hunter Boot Ltd. was born, enabling international expansion; Hunter opened a showroom on 7th Avenue, New York, and Carnaby Street in London.

2007
Launched a range of wellington boots with the Royal Horticultural Society at the Chelsea Flower Show, the start of an ongoing collaboration

2009
Hunter announced that it would be collaborating with Jimmy Choo for a limited edition black wellington boot; a second collaboration with the brand was launched in 2011

2013
Alasdhair Willis as appointed as creative director

2014
Opens first stand-alone store at Regent St., London

2014
First runway show at London Fashion Week

2014
Hunter holds an event to celebrate the launch of the Hunter Original autumn/winter 2014 collection with Urban Outfitters in the Urban Outfitters store at Herald Square, New York

2015
Launches Japanese website, hunterboots.jp, and announces plans to open a flagship store in Tokyo

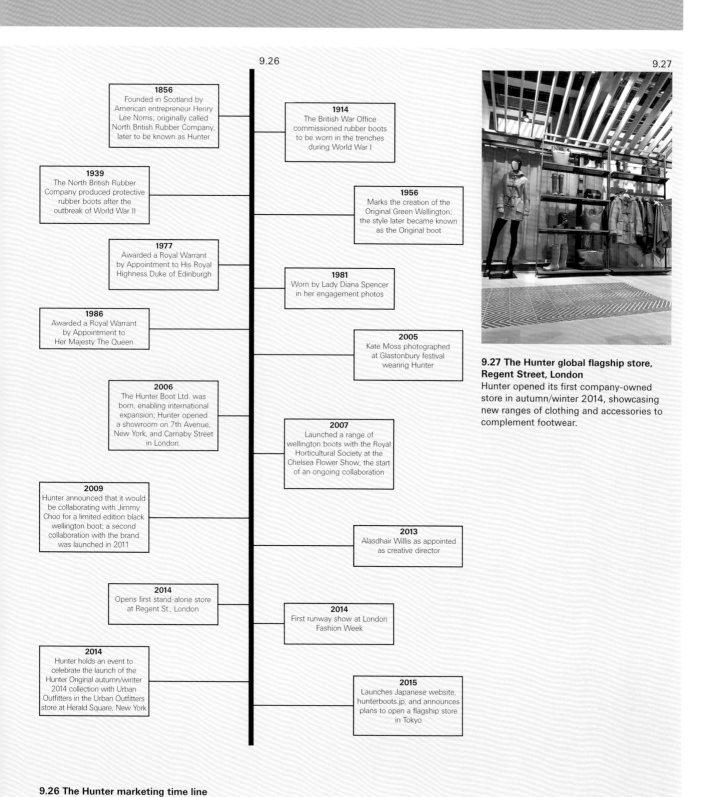

9.27

9.27 The Hunter global flagship store, Regent Street, London
Hunter opened its first company-owned store in autumn/winter 2014, showcasing new ranges of clothing and accessories to complement footwear.

9.26 The Hunter marketing time line
Key events in the brand's marketing history: the Hunter brand journey so far.

Industry Perspective:
Michelle Crowe, Email Marketing Co-ordinator at Kurt Geiger

Michelle graduated with first-class honors in BSc in International Fashion Marketing in 2010. Her role at Kurt Geiger includes the coordination of weekly trade email communications, planning, briefing internal design team, approving for HTML build, and sending out ten-plus campaigns per week. She is also responsible for the data analysis and segmentation, particularly the analysis of customer purchase and behavior data to develop strategy and segmentation plan.

Describe your typical day.

A typical day would begin by checking my email calendar, ensuring any email campaigns to send out are on track. I check the performance of the week's emails so far, looking at engagement and analyzing conversion rates. I manage the whole email marketing activity, from content planning to execution and analysis. I also spend time analyzing the customer database, looking at email segmentation strategies and identifying customer profiles.

How did you get to where you are now?

In 2010 I graduated in international fashion marketing and took an initial unpaid internship at figleaves.com, which after two weeks became paid temp work and ultimately permanent. After a year a role on the digital marketing team came up. I mainly assisted the coordination and analysis of email marketing campaigns, pay per click, and affiliate activity. As I worked across all digital marketing platforms I developed a really solid foundation knowledge of online marketing. My background in off-line marketing also helped my understanding of the advantages of online marketing versus traditional off-line methods. After 18 months I needed a new challenge, and an ambition of mine [was] to work for a major high street retailer and I wanted to also consider multichannel, both on- and off-line retail, so I moved to Kurt Geiger.

What is the best piece of advice you have been given?

Somebody once told me to not worry about money and salary but to focus on being the best you can be in your current role; if you are good at your job, the money will come. I think all graduates are keen to start climbing the career/salary ladder, and unfortunately money is often viewed as a key indicator as to how successful you are. I personally believe as long as you are challenged, you are learning and you feel your career is heading in the right direction, then salary and job titles are somewhat irrelevant.

What advice would you give to someone in college and in the early stages of his or her career?

To create opportunities for yourself and to make the most of them. Interning is a great way to see inside the real working world of fashion retail, and it is also a great way to start networking. Every interview, whether successful or not, is a great opportunity to get a look inside another business, to gain more confidence in your ability, to discover how you can improve. Forging out a successful career for yourself can sometimes feel slow and steady, but if you have the determination you will get there in the end.

What does the future hold for you and for future generations of fashion footwear employees?

In the future I definitely want to go down the customer data/CRM route. I want to understand more about the customer and their buying behaviors; I want to be able to target them and improve the relevancy of marketing. I think as technologies improve and retailers are acquiring more information about their customers, consumers will expect a more personal service.

SUMMARY

Understanding the full scope of marketing communications and how upcoming designers and established brands alike can best reach their consumer is a complex process. Advertising, public relations, events management, sponsorships, direct marketing, and retail marketing are the main conduits whereby an existing, or new, consumer can be reached. It's important not to overlook the importance of trade marketing also, because a company's reputation with key opinion leaders within the industry is also a crucial part of the overall reputation and image of a brand or designer.

While it's important that campaigns are visually impactful and compelling, if they are not part of an overall strategy, and are even at odds with a brand's or designer's business objectives, this can create both short-term and long-term issues for a business. Fully integrated campaigns are most successful when they are grounded in robust research that supports and informs the creative direction of the campaign. Digital developments in the field of marketing have changed the landscape of traditional marketing to include more information about the customer, by following both their behavior on websites and their engagement with social media. This brings with it a greater ability to engage and understand the consumer; therefore, retailers and brands should be able to target and serve the customer better.

DISCUSSION QUESTIONS

1. What are the advantages and disadvantages of measuring campaigns on ROI alone?

2. What are the benefits of product seeding?

3. How can brands best use IMCs to their advantage?

4. How are digital "trunk shows" different from the traditional format?

5. Can and should bloggers and vloggers be impartial?

EXERCISES

1. Examine the role that celebrities have played in the promotion of fashion footwear over the past decade. How may this role adapt in the future?

2. Outline the traditional tools of marketing communications used in fashion footwear. How are these being integrated into digital formats?

3. Public relations and event management are an integral part of an IMC plan used by fashion footwear companies to achieve effective marketing communications. Identify a footwear brand of your choice and develop an appropriate event and promotional campaign to support the event.

KEY TERMS

Advertising—paid-for communication from the brand that is delivered via mass media such as magazines, TV, or billboards

Advertising Standards Authority (ASA)—self-governing body regulating standards for UK advertising

AIDA (attention, interest, desire, action)—a model used to better understand customer engagement in campaigns

Apps—short for application; a computer program usually downloaded onto a mobile device to serve a particular process—e.g., shopping from a precise retailer or microblogging

Blog—a website that is regularly updated with fashion news and images, usually written by an individual or small team

Brand audit—an examination of the current status of a company's brand with a view to building or developing a strategy

Content (media)—information that provides knowledge, images, and material for the reader or viewer, i.e., the end user

Direct marketing—messages sent directly to the consumer from the brand through email, post, etc.

Editorial—published information in newspapers and magazines written and created by journalists and stylists

Events management—overseeing and orchestrating an event to maximize impact, reach, and brand awareness

Federal trade commission (FTC)—government body that regulates advertising and marketing law in the United States

Google analytics—online service provided by Google to track and report website traffic

Influencer marketing—engaging with new digital influencers such as bloggers, as well as the more traditional use of celebrities

Integrated marketing communications (IMC)—a company's marketing strategy that combines several traditional and digital promotional tools that should complement and support a successful strategy

Mass media—traditional forms of communication to a mass homogenous markets such as newspapers, TV, radio, and magazines

Micro blogging—similar to blogging, but short, succinct messages of 140 characters or images such as Twitter or Instagram

Millennials—consumers aged eighteen to thirty-four, born in the twenty years between 1982 and 2000, approximately

National Media Council—government body that regulates advertising and marketing law in the UAE

Pay per click—the advertiser pays the host website when an ad is clicked through to the brand's website

Point of purchase (POP)/point of sale (POS)—information, logos, and visuals in store or at the till point that enhance brand awareness, promotion, etc. to prompt a sale

Press release—information written in a standardized format and sent to journalists and bloggers for dissemination to the public

Product seeding—gifting upcoming and sometimes established opinion formers such as celebrities, who can support and grow the brand

Public relations (PR)—representative of the brand who manages relationships with key stakeholders such as consumers and journalists

Quick response codes (QR codes)—a matrix bar code that is scanned by a mobile phone that links directly to a brands website

Reach—total number of people that the message will spread to and potentially influence

Retail marketing—activity that related directly to the promotion of retail / in store activities

Return on investment (ROI)—an assessment of the profit made or success of a promotional tool or wider campaign used based on the costs incurred

Richness—depth and quality of information which supports the customer experience

Search engine optimization (SEO)—the process of creating website visibility organically or for free

Social media—various computer or digital-based communities and websites that allow individuals to post information and interact with each other to a global audience

Sponsorship—funding and support for an event or person by a brand that can enhance their reputation and visibility

Trade marketing—activity that relates directly to the promotion of a brand to other businesses, such as retailers or journalists, known as business to business (B2B)

Trunk shows—retail events that are often a collaboration between the brand and retailer for VIP customers

Viral marketing—activity that relates directly to the promotion of a brand or product via the internet that is passed on by third parties

Bibliography

Akehurst, G. & Alexander, N., eds. (1996). *The Internationalisation of Retailing*, London: Routledge.

Andrews, R., et al. (1998). *Economics of World Shoe Production Trends, Economics, Ethics, and the Impacts of the Global Economy: The Nike Example*, Galahad Clark. Chapel Hill: The University of North Carolina.

Bailey, S., & Baker, J. (2014). *Visual Merchandising for Fashion*, London: Fairchild Books.

Baron, S., et al. (2001). Retail theatre: The intended effect of the performance. *Journal of Service Research*, 4(2), 102–117.

Belk, R. (2003). "Shoes and Self," in NA Advances in *Consumer Research*, Vol. 30, eds. Punam D., & Rook, D., Valdosta, GA: Association for Consumer Research, 27–33.

Chandler, D. (2006). *Semiotics: The Basics*, London: Routledge.

Cheah, L. et al. (2013). Manufacturing-focused emissions reductions in footwear production. *Journal of Cleaner Production*, 44, 18–29.

Crewe, L. (2013). When virtual and material worlds collide: democratic fashion in the digital age. *Environment and Planning*, 45(4), 760–780

Davis, M. (2009). *The Fundamentals of Branding*, London: AVA Publishing SA.

Dunoff, J. & Moore, M. (2014). Footloose and duty-free? Reflections on European Union—Anti-Dumping Measures on Certain Footwear from China, *World Trade Review*, 13(2), 149–178.

Easey, M. (2009). *Fashion Marketing*, 3rd ed., Oxford: Wiley-Blackwell.

Gebre-Egziabher, T. (2007). Impacts of Chinese imports and coping strategies of local producers: the case of small-scale footwear enterprises in Ethiopia. *Journal of Modern African Studies*, 45(4), 647–679.

Grose, V. (2012). *Concept to Customer*, London: AVA Publishing SA.

Hofstede, G. (1984). *Culture's Consequences: International Differences in Work-Related Values*, London: Sage.

Hsing, Y. (1999). Trading companies in Taiwan's fashion shoe networks. *Journal of International Economics*, 48, 101–120.

Jackson, T. & Shaw, D. (2009). *Mastering Fashion Marketing*, London: Palgrave Macmillan.

Jackson, T. & Shaw, D. (2001). *Mastering Fashion Buying and Merchandise Management*, London: Palgrave Macmillan.

Jimenez, G. & Kolsun, B. (2014). *Fashion Law*, London: Fairchild Books.

Jones, R. (2006). *The Apparel Industry*. 2nd ed., Oxford: Wiley-Blackwell.

Kapferer, J. & Bastien, V. (2012). *The Luxury Strategy: Break the Rules of Marketing to Build Luxury Brands*, London: Kogan Page Ltd.

Keller, K. (2013). *Strategic Brand Management: Building, Measuring, and Managing Brand Equity,* Harlow, Essex: Pearson Education Limited.

Keller, K. (1993). Conceptualizing, measuring, and managing customer-based brand equity. *Journal of Marketing*, 57(1) 1–22 Published by American Marketing Association.

Kincade, D. & Gibson, F. (2010). *Merchandising of Fashion Products*, Harlow, Essex: Pearson Education Limited.

Kotler, P., et al. (2009). *Marketing Management*, Harlow, Essex: Pearson Education Limited.

Lea-Greenwood, G. (2013). *Fashion Marketing Communications*, Oxford: Wiley-Blackwell.

Mehta, R., et al. (2010). Managing international distribution channel partners: A cross-cultural approach. *Journal of Marketing Channels*, 17, 89–117.

Mesher, L. (2010). *Retail Design*, London: AVA Publishing SA.

Miles, R. E., et al. (1978). Organizational strategy, structure, and process. *Academy of Management Review*, 546–562.

Mintel (2015). Footwear — US, Mintel Market Sizes.

Moore, J. (2009) V*isions of Culture: An Introduction to Anthropological Theories and Theorists* (Third Edition). Lanham and New York: Alta Mira Press, 141.

Nice Classification. (2016). 10th ed.

Ogilvy, D. (1983). *Ogilvy on Advertising*. Toronto: John Wiley and Sons.

Ojeda-Gomez J., et al. (2007). Achieving competitive advantage in the Mexican footwear industry. *Benchmarking: An International Journal*, 14(3), 289–305.

Park, T. & Curwen, L. G. (2013). No pain, no gain? Dissatisfied female consumers' anecdotes with footwear products. *International Journal of Fashion Design, Technology and Education*, 6(1).

Porter, M. (2004). *Competitive Advantage*, New edition, New York: Free Press.

Posner, H. (2011). *Marketing Fashion*, London: Laurence King Publishing Ltd.

Rabolt, N. & Miller, J. (2009). *Concepts and Cases in Retail and Merchandise Management*, 2nd ed., London: Fairchild Books.

Raymond, M. (2010). *Trend Forecasters Handbook*, London: Laurence King Publishing Ltd.

Rexford N. E. (2000). *Women's Shoes in America 1795–1930*, The Kent State University Press.

Riello, G. & McNeil, P. (2011). *Shoes: A History from Sandals to Sneakers*, Oxford: Berg.

Rinallo, D. & Basuroy, S. (2009). Does advertising spending influence media coverage of the advertiser? *Journal of Marketing*, (73)6, 33–46.

Roach, M. (2003). *Dr. Martens—The Story of an Icon*, Chrysalis Impact.

Rogers, E. (2003). *Diffusion of Innovations*, 5th ed., New York: Free Press.

Shaw, D. & Koumbis, D. (2014). *Fashion Buying: From Trend Forecasting to Shop Floor*, London: Fairchild Books.

Solomon, M. & Rabolt, N. (2009). *Consumer Behavior in Fashion*, 2nd ed., Harlow, Essex: Pearson Education Limited.

Staikosa, T. & Rahimifarda, S. (2007). A decision-making model for waste management in the footwear industry. *International Journal of Production Research*, 45(18–19).

Tristão H., et al. (2013). Innovation in industrial clusters: a survey of footwear companies in Brazil. *Journal of Technology Management & Innovation*, 8(3), 45–56.

Walford, J. (2010). *Shoes A–Z— Designers, Brands, Manufacturers and Retailers*, London: Thames and Hudson.

Walker, H. (2012). *Cult Shoes-Classic and Contemporary Designs*, Merrell Publishers.

World Footwear Yearbook (2015).

Yoon Kin Tong, D., et al. (2012). Ladies' purchase intention during retail shoes sales promotions. *International Journal of Retail & Distribution Management*, 40(2), 90–108.

Further Resources

KEY INTERNATIONAL FASHION FOOTWEAR TRADE FAIRS

Country	Event	Location	Website
Brazil	Francal	Sao Paulo	http://www.francal.com.br/site/
China	Pure	Shanghai	http://www.pureshanghaishow.com/
	Fashion Access	Hong Kong	http://www.fashionaccess.aplf.com/
	Shoes and Leather	Guangzhou	http://www.shoesleather-guangzhou.com/home.html
Denmark	CPH	Copenhagen	http://www.copenhagenfashionweek.com
France	Who's Next	Paris	http://www.whosnext.com
	Première Classe	Paris	http://www.premiere-classe.com/en/
	Tranoï Femme	Paris	http://www.tranoi.com
	Paris Fashion Week	Paris	http://www.modeaparis.com/en
Germany	GDS	Dusseldorf	http://www.gds-online.com
	Premium	Berlin	http://www.premiumexhibitions.com
	Seek	Berlin	http://www.seekexhibitions.com
Italy	Pitti	Florence	http://www.pittimmagine.com/en/corporate.html
	Micam	Milan	http://www.micamonline.com
	Milan Fashion Week	Milan	http://www.cameramoda.it/en/
UK	London Fashion Week	London	http://www.londonfashionweek.co.uk
	Pure	London	http://www.purelondon.com
	Scoop	London	http://www.scoop-international.com
	MODA	Birmingham	http://www.moda-uk.co.uk
USA	New York Fashion Week	New York City	http://www.mbfashionweek.com
	FFANY, New York,	New York City	http://www.ffany.org
	Magic	Las Vegas	http://www.magiconline.com
	FN Platform	Las Vegas	http://www.magiconline.com/fn-platform
	The Atlanta Shoe Market	Atlanta	http://www.atlantashoemarket.com
	Transit	Los Angeles	http://www.californiamarketcenter.com/transit/brands.php
	Sole Commerce	New York City	http://www.enkshows.com/sole/

Further information at http://www.tradefairdates.com/Shoe-and-Footwear-Fairs-WB307.html

FOOTWEAR ASSOCIATIONS

Country	Association	Website
Brazil	Brazilian Association of the Footwear Industry	http://www.brazilianfootwear.com
France	Federation Francaise de la Chaussure	http://www.chaussuredefrance.com/sites/fr/index.html
Germany	The German Federal Association of the Footwear and Leather Goods Industry	http://www.hdsl.eu/Home-en
India	Indian Footwear Components Manufacturers Association	http://www.ifcoma.org
India	Indian Shoe Federation	http://www.indianshoefederation.in
Italy	Italian Manufacturers Footwear Association	http://www.assocalzaturifici.it
Poland	Polish Chamber of Shoe and Leather Industry	http://www.pips.pl
Portugal	APICCAPS—Portuguese Footwear, Components and Leather Goods Manufacturer's Association	http://www.apiccaps.pt
Spain	Federation of Spanish Footwear Industries	http://www.fice.es
UK	British Footwear Association	http://www.britishfootwearassociation.co.uk
USA	American Apparel and Footwear Association	http://www.wewear.org/
USA	Fashion Footwear Association of New York	http://www.ffany.org/

Further international listings for manufacturing and retail associations at
http://www.worldfootwear.com/organizations.asp?Organizations

WEBSITES

http://www.acas.org.uk
http://www.asa.org.uk
http://www.business.ftc.gov
http://www.drapersonline.com
http://www.footwearinsight.com
http://www.footwearnews.com
http://www.footweartoday.co.uk
http://www.international-footwear-foundation.co.uk
http://www.shoeinfonet.com
http://www.shoetrades.com
http://www.wipo.int
http://www.worldfootwear.com

Acknowledgments and Picture Credits

From Fiona

Special thanks to the retailers, brands, and associations that agreed to be part of this book. We were warmly welcomed into factories, showrooms, trade fairs, and stores during our research. Without you this book would not have been possible.

Thank you to the industry individuals and academic reviewers for their overwhelming support, insight, and encouragement, which we hope has accurately helped us to fill a gap in the literature.

To our colleagues, the photographers, and illustrators, we are in your debt.

To Mark and William, Mum and Dad—thank you.

From Tamsin

Thank you to our commissioning editor Colette Meacher for the vision in believing in this book from the offset, to our development editor Lucy Tipton for your professionalism and tenacity to see this project through to its completion, and to Helen Stallion for your positivity and help with the images.

Thanks to Southampton Solent University Research and Enterprise panel and friends, students, and colleagues for all your help and encouragement. And to everyone from the industry who gave their time and knowledge generously.

And huge thanks to my husband and Mum for your love, patience, and support.

Picture credits

1.1, 1.10, 1.12–16, 1.23, 2.1, 2.3, 2.8–2.12, 2.24–2.25, 2.27–2.28, 3.1, 3.5, 3.13, 4.2, 5.9, 5.11–5.12, 6.7, 9.2, 9.15, 9.16, 9.20 Fiona Armstrong-Gibbs; 1.3 KC Photography/Getty Images; 1.4 Photography: Andree Martis and Demian Dupuis; Makeup: Porsche Poon; Hair: Roger Cho; Model: Soraya Jansen@NEVS, Jakub Zgid; Graphic Post Production: Jakub Zgid; Textile Collaboration: Demian Huang/Fukurou Living; Footwear photography: Jakub Zgid; 1.8 Kirstin Sinclair/Getty Images; 1.9 Julia Davila-Lampe/Getty Images; 1.11 Ocean Sole/http://www.ocean-sole.com; 1.17 Jupiterimages/Getty Images; 1.19 Antonio Berardi; 1.21 Photo by Jason Merritt/Getty Images; 1.24–1.25 The Boot Tree Ltd.; 1.26 Photo by Kirstin Sinclair/Getty Images; 1.27 The Boot Tree Ltd. and Sue Barr (photographer); 1.28–1.29 Jason Fulton; Portrait of Jason Fulton: Atlynn Vrolijk; 2.4 Andy Freeney ARPS; 2.6 Hotter; 2.7 Andy Freeney; 2.14 (Photo) johnnyscriv/Getty Images; 2.15 Harvey Nichols; 2.16 Oka-B/photograph by Anthony Kitchens; 2.17 Harvey Nichols; 2.18 Jakub Zgid; 2.21 Loake; 2.22 Oka-B/photograph by Anthony Kitchens; 2.23 Photo by Antonio RIBEIRO/Gamma-Rapho via Getty Images; 2.26 MARK RALSTON/AFP/Getty Images; 2.29–2.31 Soft Star; 3.7 Harvey Nichols; 3.6 Photo by Raphael Huenerfauth/Photothek via Getty Images; 3.12 Sarah Fuller; 3.14–3.15, 4.1, 4.4, 4.7, 4.8, 4.11–4.15, 6.1, 6.3, 6.4, 6.8–6.16, 6.18, 6.19, 7.5–7.10, 7.12, 7.15–7.17, 8.2, 8.7–8.9, 9.7, 9.26 Tamsin McLaren; Portrait of John Saunders and bfa logo: by permission of John Saunders; 4.10 Farfetch; 4.16 Brian Armstrong; 4.17 and portrait of Mary Stuart: Mary Stuart; 5.1 Robin Clewley; 5.4 Photo by Kevin Mazur/Getty Images for 42West; 5.6– 5.7 The Boot Tree Ltd. and Sue Barr (photographer); 5.13–5.15 Tanya Heath; 6.2 Designed by Four-by-Two, photography by Chris Humphreys Photography; 6.6 Saks Fifth Avenue/Russell Howe; 6.17 Farfetch; 6.20–6.21 THE ONE OFF; Portrait of Marc Debieux: Sam Jackson; 7.1 Jerome Favre/Bloomberg via Getty Images; 7.2 Photo by Nicholas Devore III/National Geographic/Getty Images; 7.14 Photo by Joe Scarnici/WireImage/Getty Images; 8.1 Gavin Watson; 8.4 Photo by Jamie McCarthy/WireImage for Jenni Kayne/Getty Images; 8.5–8.6 Amy Cox and Kat Maconie; 8.10–8.11 Gandys; 8.12 Gavin Watson; 8.13–8.14 James Pearson; 8.15 Moni Haworth; 8.16 and portrait of Tracey Neuls: Tracey Neuls; 9.1 Photo by Dave Hogan/Getty Images; 9.4 Mike Harrington/Getty Images; 9.5–9.6 InStyle; 9.8 William Anthony Kitchens; 9.9 Photo by Kirstin Sinclair/FilmMagic; 9.10 Photo by Ian Gavan/Getty Images; 9.11 Photo by George Pimentel/WireImage/Getty Images; 9.13 Kit Houghton; 9.14 Photo by Suzanne Kreiter/The Boston Globe via Getty Images; 9.17 Photo by Kirstin Sinclair/Getty Images; 9.18 Photo by Vittorio Zunino Celotto/Getty Images; 9.19 William Anthony Kitchens; 9.21 Amy La; 9.22 Yosi Samra Inc; 9.23 Harvey Nichols Dubai; 9.24 Photo by Ian Gavan/Getty Images